D0992385

The Mosses
of Michigan

The Mosses

of Michigan

HENRY T. DARLINGTON

Cranbrook Institute of Science

BULLETIN NO. 47 • 1964

Edited by Howard Crum

Published with the aid of
The Edwin S. George Publication Fund

Designed by William Bostick
Set in type and printed by The Cranbrook Press.
Offset plates by The Meriden Gravure Co.
Bound by The Kingsport Press.

To the memory of

MY MOTHER

Jane Sellers Darlington

A friend of all wild life

Preface

Mosses are abundantly represented throughout the world except in the most extreme of cold or arid regions. Visitors to the mountains or the more or less virgin forests of the northern states can hardly fail to be impressed by the abundance and beauty of the mosses. The green surfaces of old logs, rocks, and tree bases in almost any wooded area largely result from the presence of mosses or the closely related liverworts. Aside from their value in the adornment of nature, mosses play an important role in soil conservation and plant succession. The spongy mats or sods of many species lessen the forces of erosion by preventing the rapid run-off of flood waters. In the far North a deep permafrost results not only from the extreme cold winter but also from the thick blanket of mosses and lichens which prevents thawing in the summer. The poor drainage which makes much of the North virtually inaccessible except by air results in part from the clogging and damming of streams by bog vegetation, which is largely mossy. As pioneers on bare soil and rock, many mosses help initiate the vegetative succession leading eventually to the climax of the region and the mature soils associated with the climax. In past ages, mosses have made significant contributions to the formation of the world's peat deposits. Some practical values, such as the extensive use of peat in gardening or as fuel or the utilization of undecomposed peat moss in shipping nursery stock or crockery, might also be mentioned. During the First World War, *Sphagnum* (better known as peat moss) was used in surgical dressings because of its absorptive and naturally sterile character.

To the casual observer the "moss plant" is the green clump or mat; it is often not realized that each clump usually contains many individual plants. The illustrations in this manual show single plants for many species. Here it is seen that the parts of the individual plant usually visible and most striking are the leafy stem and the stalk with the spore-bearing portion at its tip. Not all mosses are conspicuously green, and less noticeable kinds often grow in more or less open situations, on the ground, on dry, exposed rocks or ledges, in open fields, and along roadsides.

The true mosses (Musci) form the greater part of the division Bryophyta; the lesser portion are the Hepaticae. Only the true mosses are treated in this work. The majority are easily distinguished by their general appearance from lichens with which they are sometimes confused and which frequently grow in similar habitats. Lichens are usually gray or grayish-green, especially when dry. The more common or conspicuous forms may often be seen on the bark of trees as more or less wrinkled crusts. Others are more noticeable on the ground in the open. An example is the "reindeer moss," which may cover large areas in arctic regions. It is a freely branched, bushy type; most lichens, however, grow as flat expansions on which small, disc-shaped fruiting structures are often present. The body of a lichen (thallus) never bears leafy shoots as in a moss. Lichens, which consist of algae and fungi growing together, belong to a lower division of the plant kingdom and have a structure quite different from that of mosses.

The moss flora of Michigan is best known through the work of Dr. George E. Nichols and Dr. William C. Steere, who published extensively on their collections from 1920 to 1942. During the winter of 1936-37, they jointly prepared a preliminary list of Michigan bryophytes, which Dr. Nichols made available in mimeographed form to his students at the University of Michigan Biological Station. This list contains, besides the Hepaticae, 18 *Sphagna* and 351 Musci. Owing to the death of Dr. Nichols in 1939, this catalogue was not published. By 1942 a number of species (including 10 Musci) had been added to the known bryophyte flora of the state, but, because of the war, completion of the catalogue by Dr. Steere was postponed and publication delayed until 1947. Since that time a few more species and at least three more genera have been reported. However, only a few of the 83 counties of the state have been intensively explored for bryophytes. Casual collecting has been done in about 50 counties; for about 25 counties there is not a single record. The time now seems ripe for a renaissance of exploration and for annotations on the distribution of species in the state.

This manual has been prepared as a guide to the moss flora of the state insofar as it is known. For the more common species, it will undoubtedly have considerable coverage for neighboring states in the Great Lakes region. Up to the present there have been reported 43 families, 125 genera, 377 species, and 30 varieties. While this manual provides a record of the known moss flora of the state and its distribution, the author realizes that it can never be complete. New species and even genera are being discovered from time to time. From more general works, the range of other species has been reported for the Michigan area.

In the introduction the writer has outlined the history of bryological investigations in the state and also the salient features of topography and climate that have resulted in a diversity of species. The moss flora of the Upper Peninsula has certain features which contrast strongly with those of the Lower Peninsula. For that reason, the writer has adopted Dr. Steere's suggestion of keeping the records for the two peninsulas separate.

The genera and species and their arrangement and nomenclature are based largely on Steere's catalogue of 1947. In a few cases, generic names which occur more commonly in American publications have been adopted. The key to the genera of the Bryales, beginning on page 12, forms the basic approach to the moss flora of the state. In the construction of this key the writer has to some extent been guided by the outstanding characters exhibited by the local representatives of the genera. The keys to both genera and species are dichotomous. Emphasis has been placed on field characters whenever they are at all distinctive.

In the preparation of this volume, the beginner of the study of mosses has not been forgotten. It is believed that the keys, together with the descriptive material covering the families, genera and species, while lacking the detail to be found in more general works, will meet the needs of those who have not had the advantage of a course in bryology at a university or biological station. The key to the genera of the order Bryales, beginning on page 12, has been prepared by the writer primarily for the use of those beginning the study of the mosses.

As Figure 2 will be found a chart showing the relative length of various cells to their breadth. This is particularly useful in connection with the Pleurocarpi. The writer has found such a chart helpful in his own work.

On page 10 will be found suggestions as to necessary equipment for the study of mosses and also methods of collection and preservation of specimens. The life history of a moss plant will be found on page 9.

The collections recorded for each species have come from various sources. The county has been used as the unit of distribution, but more specific localities are sometimes mentioned. Distribution records, of course, can never be complete; many of the more common species probably occur in every county of the state. However, the listings for such species should be of interest to beginners. While there would be no purpose in a manual of this character in extending such records beyond a reasonable length, the ecologist or phytogeographer interested in variations of habitat, relative distribution, or range might desire an even more complete set of records. A list of the collectors cited in this work will be found on page 16.

The writer accepts all responsibility for the choice of topics discussed in the non-technical portion of the manual. The separation of the mosses into the Acrocarpi and the Pleurocarpi, which has been accepted by authors for many years, has certain advantages that the writer wishes to retain in the general key.

The distribution records are rather unique in containing not only the names of visiting bryologists but also those of many local collectors; and in citing works which some moss collectors may already have or which may be available in local libraries. Synonymy has been omitted except where it was considered important.

For a very important contribution to this manual, the writer is deeply indebted to Dr. Irma Schnooberger, who prepared the key to the difficult genus *Sphagnum*, with the accompanying descriptions and illustrations, and in addition furnished distribution records for a number of the species. Dr. A. Leroy Andrews kindly read the descriptions of the species of *Sphagnum*. Dr. Winona H. Welch checked the key to the Michigan species of *Fontinalis* and added a number of species previously unrecorded for the state. Dr. Geneva Sayre checked the key to the difficult genus *Grimmia*. To these bryologists the author wishes to express sincere thanks. The writer is also deeply indebted to Dr. A. J. Sharp for reading the keys and descriptions.

My greatest debt is to Dr. Howard Crum, of the National Museum of Canada, who critically and carefully edited the entire manuscript. Assisting Dr. Crum, Miss Hilda T. Harkness gave generously of time and talent.

The writer owes much to Dr. W. C. Steere not only for his encouragement in the collection of mosses and his determinations of the many specimens submitted to him throughout the years, but also for his encouragement in the preparation of this manual and for reading the manuscript. The preparation of this manual has entailed a certain amount of expense for travel and especially for clerical work; aid for these expenses has been furnished largely by an All University Research Grant. In this connection the writer deeply appreciates the interest and helpfulness extended by Dr. W. B. Drew, chairman of the Department of Botany and Plant Pathology at Michigan State University. To Dr. Robert T. Hatt I am indeed grateful for his finding the means to publish this manuscript and for his labors in seeing it through production.

It is too much to expect this book to lack errors and omissions. Where shortcomings of any nature occur, the writer will appreciate having them pointed out.

H. T. DARLINGTON

October, 1963

Table of Contents

County map faces page 1
Illustrations follow page 84

THE COUNTIES OF
MICHIGAN

Introduction

BRYOLOGICAL EXPLORATION OF MICHIGAN

The early collecting and recording of the plants of Michigan were naturally done in connection with the vascular species. Dennis Cooley, a physician and botanist living at Cooley's Corners (now the village of Washington) in Macomb County, was apparently one of the first to collect mosses. His specimens are now in the herbarium of Michigan State University. Some of his labels show that he corresponded with William Sullivant, who must have identified the specimens. Though Cooley collected vascular plants in Michigan as early as 1827, his bryophyte collections were apparently made from about 1848 to 1852.

Toward the close of the century Charles F. Wheeler and E. J. Hill were collecting mosses in the state. Hill (1885) mentioned finding *Mnium serratum* on a visit to the Menominee Iron Region. Later he reported finding *Fissidens grandifrons* (1902) and *Amblystegium noterophilum* (1909) at Boyne Falls in Charlevoix County. Wheeler, of the Michigan Agricultural College, traveled widely in the Lower and Upper Peninsulas. He was mainly interested in vascular plants but also collected bryophytes from about 1890 to 1900 in at least three counties in the Upper Peninsula (Alger, Delta, and Dickinson counties) and seven in the Lower Peninsula (Alpena, Cheboygan, Clinton, Eaton, Emmet, Ingham, and St. Joseph counties). His total findings of Musci, numbering about 70 species, were checked or determined by W. C. Steere in 1933; all are deposited in the Michigan State University Herbarium.

A list of mosses based on a collection made by A. Purpus in Jackson County in 1891 appeared in an article by J. Röll in 1897; this appears to have been the first published list of Michigan mosses. The period beginning in 1902 and ending about 1920 may be considered as introductory to a more extensive exploration of the bryophyte flora of the state. During this period ecological research on Isle Royale resulted in a rather definite knowledge of the bryophytes in a limited area; C. E. Allen and S. C. Stuntz spent five weeks on the island in 1901 and collected 36 species of liverworts and several species of mosses. In 1905, W. P. Holt, as a member of the biological survey of Isle Royale under A. G. Ruthven, collected 38 species of Musci (Holt, 1909). The most extensive addition to our knowledge of the mosses of Isle Royale during this period was made by W. S. Cooper, who spent the summers of 1909 and 1910 determining the vegetational climax of the island and published a paper entitled

"A list of mosses collected upon Isle Royale, Lake Superior" (1913). In the summer of 1914, A. H. Povah, as cryptogamic botanist of the Shiras Expedition to the Whitefish Point region of Chippewa County, collected 26 species of Bryales and three species of *Sphagnum*. His findings, consisting mostly of fungi and a few algae, were reported 15 years later (1929) under the title "Some non-vascular cryptogams from Vermilion, Chippewa County, Michigan." C. H. Kauffman had already published a paper in 1915 entitled "A preliminary list of the bryophytes of Michigan," in which he listed 105 species of Bryales collected by himself and his students in the southern part of the state. The majority of the specimens were collected in the vicinity of Ann Arbor, though numerous references are made in his paper to Alpena and Holland, points frequently visited in the prosecution of his work on the fungi. In his paper, no mention is made of *Sphagnum;* apparently the first important collections of this genus were made at the Douglas Lake biological station in 1918 and 1919 by William Praeger of Kalamazoo College (Praeger, 1919).

The most active period in the exploration of the bryophyte flora began in 1920, when G. E. Nichols, a well-known ecologist and bryologist from Yale University, joined the staff of the University of Michigan Biological Station at Douglas Lake in Cheboygan County. During his first 12 years at the station, Nichols collected mainly in the vicinity of the station, at Burt Lake nearby, in neighboring Emmet County, and on Mackinac Island. His first report, published in 1922, listed 321 species of bryophytes for the state. Two additional reports, in 1925 and 1932, increased the number of species known for the state to 366. Of the additional species in the 1932 report, some of the most noteworthy were discovered by H. R. Becker, a keen observer living at Climax, Kalamazoo County; he detected at least 12 species in that region not previously known to occur in the state. A consideration of this fact emphasizes the importance of an intensive survey of even a limited area.

When Nichols began his work at the station, only four papers had been devoted to the mosses of the state, the two most extensive being by Cooper on the mosses of Isle Royale and by Kauffman on scattered collections mostly in the Lower Peninsula. The major portion of the Upper Peninsula was little known bryologically; however, some of the unusual rock formations of that region seemed to merit examination for species not recorded elsewhere from the state. It was not until the summer of 1933 that Nichols began serious collecting in the Upper Peninsula. Accompanied by Carol Lander, F. W. Tinney, and H. A. Gleason, he made a two-day trip to Miner's Castle and Miner's Falls in the Pictured Rocks region and to the region west of Munising, all in Alger County (Nichols, 1933). The discovery of more than 25 rock-inhabiting liverworts and mosses (including those collected by W. C. Steere during the same season) emphasized the bryological importance of the region. The following summer (1934), Nichols continued his explorations in Alger County and during a part of the time spent a week in the mountains of Marquette County as a guest of the Huron Mountain Club. Here he was rewarded by the discovery of several rock-inhabiting species either new to the state or found previously only in the Munising region (Nichols, 1935). These species gave unmistakable evidence of an arctic-alpine element in the bryophyte flora of the state.

In the summer of 1935, Nichols and Steere spent a week collecting in the Porcupine Mountains in Ontonagon County and prepared a list (1937) of 219 bryophytes from this rugged region extending along the south shore of Lake Superior. Nichols collected later in the same season at Tahquamenon Falls in Luce County and at scattered points in Marquette and Cheboygan counties. Up to a year before his death in 1939, he continued his studies on the bryophyte flora of Cheboygan, Emmet, and Presque Isle counties.

The years 1933-35 marked a period when much was written about the bryophytes of the Upper Peninsula. Actively engaged in this exploratory work was W. C. Steere, who had collected bryophytes around Ann Arbor since 1926. In 1931 he published "Notes on the mosses of southern Michigan" and in 1933 a supplementary report under the same title. These lists included quite a number of species unknown for the region before that time and not included in Kauffman's list published in 1915. Steere began his bryological explorations in the Upper Peninsula in 1933. In June of that year, he collected nearly 150 species on Sugar Island, a portion of Chippewa County lying in the St. Mary's River near the Ontario boundary (Steere, 1934). Later in the summer he visited the more rocky formations in the northern part of the Upper Peninsula, first collecting along the sandstone cliffs and accessible waterfalls east and west of Munising in Alger County and then in the comparatively drier, granitic rocks of the Keweenaw Peninsula. Near the end of the summer, accompanied by H. R. Becker, he visited the low sandstone cliffs along the Grand River near Grand Ledge, Eaton County; this area has yielded a number of plants not found in the more uniform terrain of the surrounding region. In the summer of 1934, he again visited the sandstone cliffs of northern Alger County and explored some high limestone formations facing Lake Michigan along the shore of Delta County. His joint collecting with Nichols the next year in the Porcupine Mountains furnished him an opportunity to make some independent collections in the same general region. In September 1936, he continued his investigations in the Keweenaw

Peninsula; his explorations were centered along the coastline from Eagle Harbor south to Horseshoe Harbor on the west side of the Peninsula and along the eastern shore in the vicinity of Jacobsville and Bête Grise, as well as in much of the more accessible inland portions of the county. Evidence had been accumulating, supported by a paper by M. L. Fernald on the disjunct occurrence in the region of certain vascular plants indigenous farther west, that a number of disjunct bryophytes might be found in the supposedly unglaciated areas of this region. Steere's intensive collecting disclosed a number of Cordilleran, Pacific, and arctic species of bryophytes. His reports on these findings were published in 1937 and later in 1938 under the title "Critical bryophytes from the Keweenaw Peninsula, Michigan." The first paper, appearing in *Rhodora,* is a masterly discussion of the bryophyte flora of the region; the 1938 report, listing some additional species of particular interest, was published in *Annales Bryologici.* In 1939, Steere joined the staff of the University of Michigan Biological Station, with which he was associated for a number of years. His "Notes on Michigan bryophytes, IV" was published in 1942. This paper discusses the occurrence and range of some of the more noteworthy species found mostly by himself and Nichols from 1930 to 1940. The total number of bryophytes recorded up to this time was 514, including 361 Musci, 132 Hepaticae, and 21 *Sphagna.* These were summarized in "The bryophyte flora of Michigan" (1947), in which Steere included the records from a few other papers published between 1930 and 1940. One of these was "The bryophytes of Isle Royale, Lake Superior" by Frances Thorpe and A. H. Povah in 1935; the mosses and hepatics in this paper had been collected by Povah and J. L. Lowe.

The writer's interest in Michigan mosses began about 1920 and led to a desire to know the names of the common species in the vicinity of East Lansing. Having secured a volume of Elizabeth Dunham's *How to Know the Mosses* (ed. 1), he was able, in a large measure, to satisfy this curiosity. In 1922 and 1923, collections were made in the Porcupine Mountains of the Upper Peninsula and duplicates sent to A. J. Grout and R. S. Williams. From this time until about 1937, the bulk of his moss collections were made near the Kellogg Biological Station in Kalamazoo County or near the writer's cottage on Glen Lake in Leelanau County. In 1938, he reported 131 species in a paper entitled "Bryophytes collected in the vicinity of Glen Lake, Leelanau County, Michigan." Little collecting was done during the war years. In 1954, he had an opportunity to make a bryological survey of a number of counties in the southern part of the Lower Peninsula and later to examine collections made by others in counties farther north. Duplicate specimens of these have been deposited in the herbaria of Michigan State University and the University of Michigan.

Possibly no one has collected more widely in that part of the Lower Peninsula lying mostly north of the Grand River than Irma Schnooberger; her "Notes on bryophytes of central Michigan" (1940), in which 40 of the less common species are mentioned, includes four species not previously recorded for the state.

Since 1940, scattered collections of bryophytes have been made in a number of counties, particularly by George W. Parmelee. In 1946, he collected about 100 species of Musci, the majority in Ingham, Clinton, and Eaton counties, mainly within ten miles of East Lansing. In the summer of 1949, he and Charles Gilly collected bryophytes in ten counties in the Lower and seven in the Upper Peninsula on an expedition devoted primarily to the collection of vascular plants. Parmelee was stationed at Houghton in the school year of 1950–51, during which time he collected 3 species, mainly in Houghton, Keweenaw, and Baraga counties. Neither the collections made by Parmelee nor those made by Darlington since 1950 in Ingham, Eaton, and Clinton counties and in some of the extreme southern counties have been reported in the literature.

Other articles on the mosses of the state have appeared since 1955, either listing new distribution records or noting the discovery of genera or species which had not been recorded. In 1956, Harold Robinson and James Wells published a paper entitled "The bryophytes of certain limestone sinks in Alpena County, Michigan;" the sink holes, containing an unusually rich bryophyte flora, yielded 110 species, of which three had not previously been reported for the state. Other recent articles on new discoveries have been contributed by Sharp (1956), Chopra and Sharp (1961), Griffin (1962), and Mazzer and Sharp (1963). In 1962, F. J. Hermann published a list of mosses collected in Keweenaw County. Crum (1964) published an annotated list of species of counties bordering the Straits of Mackinac (the area of the University of Michigan Biological Station).

COLLECTORS

The localities cited should interest both the beginner and the experienced collector. The former gains some idea of the relative occurrence of the species identified and gains satisfaction if the plant has not been recorded for the locality or if it is apparently rare; the latter usually is more interested in finding where rare species have been found, the character of the substratum, and the range.

While the Upper and Lower Peninsulas have many species in common, the records of the two areas have been kept separate on account of the distinctive character of the geological formations in the northern part, with its abundant rocky outcrops that harbor many species not found in the Lower Peninsula.

In only a few of the 68 counties of the Lower Peninsula has extensive collecting been done or lists of species published; collecting in most of the other counties has been sporadic or entirely lacking. In the southern part of the peninsula, rather intensive collecting over a number of years has been done in Washtenaw, Kalamazoo, Clinton, and Ingham counties and in the northern part in Cheboygan, Emmet, Presque Isle, and Leelanau counties. In the 15 counties of the Upper Peninsula, parts of seven counties, including Chippewa, Alger, Mackinac, Marquette, Keweenaw, Ontonagon, and Houghton, have been carefully explored. Following is a list of collectors cited by surname in the text:

Allen, C. E.
Andrews, A. L.
Bailey, T.
Barlow, B.
Beach, Harriet
Beal, W. J.
Beattie, J. D.
Beck, E. C. G.
Becker, H. R.
Bessey, E. A.
Brown, E.
Cantlon, J. E.
Conard, H. S.
Conard, Rebecca
Cooley, D.
Cooper, W. S.
Crum, H. A.
Darlington, H. T.
Davis, C. A.
Dodge, C. K.
Ehlers, J. H.
Fulford, Margaret
Gillis, W.
Gillman, H.
Gilly, C. L.
Gleason, H. A.
Gleason, H. A., Jr.
Golley, F. B.
Grassl, C. O.
Hall, M. T.
Hermann, F. J.
Hill, E. J.
Holt, W. P.
Hyypio, P. A.
Ikenberry, G. J.
Janson, V.

Kauffman, C. H.
Knobloch, I. W.
Lander, Carol
Lowe, J. L.
Lowry, R. J.
Mains, E. B.
Marshall, T. H.
McVaugh, R.
Moss, C. E.
Mugford, E. G.
Nichols, G. E.
Osborn, C. S.
Parmelee, G. W.
Paul, Mrs. W.
Phillips, E. A.
Phinney, H. K.
Povah, A. H.
Praeger, W. E.
Purpus, A.
Roberts, C. M.
Robinson, H.
Schnooberger, Irma
Sharp, A. J.
Smith, A. H.
Squiers, T. L.
Stearns, F. L.
Steere, W. C.
Stuntz, S. C.
Tarzwell, C. M.
Thompson, Margaret
Thorpe, Frances
Tinney, F. W.
Welch, Winona H.
Wells, J.
Wheeler, C. F.
Wynne, Frances E.

THE MOSS FLORA

In an area the size of Michigan, consisting of 57,430 square miles, a great variety of habitats results in a comparatively rich moss flora. The Bryales contain by far the greatest number of species; the other two orders, the Sphagnales and the Andreaeales, form only about six percent of the moss flora. In size of colonies and far-flung distribution, the bog-inhabiting genus *Sphagnum*, with 22 recorded species, exceeds those of any other genus reported for the state. This is not surprising when we consider that there are over 11,000 lakes in the state, besides an abundance of swampy depressions which are wet at least part of the year. In addition to the common occurrence of the genus on boggy ground along the borders of many lakes, it may be found sometimes on wet rocks and moist banks. *Sphagnum* is sometimes called peat moss, though other bog mosses also enter into the formation of peat. The wet habitat, general appearance, soft texture, and usually mass habit of growth make this genus easy to recognize; however, in the determination of species, the genus is one of the most difficult. The normally upright stems are usually massed in hummocks whose upper surface consists of younger, lighter-colored growth; the lower portions on dying gradually add their remains to form beds of peat. From the stems arise numerous branches which are arranged in fascicles; these become more separated as the stem continues growth. The small, ovoid or ellipsoid leaves are adapted for absorbing and holding water like a sponge. This fact has given *Sphagnum* considerable economic value for the shipment of living plant material.

The order Andreaeales, represented by the single genus *Andreaea*, is limited in Michigan to the most northerly counties of the Upper Peninsula. It grows on siliceous rocks in small, dark, dense patches resembling those of the genus *Grimmia*, which often occurs in similar situations. However, the capsules of *Andreaea* open by four vertical slits, whereas those of *Grimmia* open by a lid; moreover, the leaves of the latter are often hair-pointed. Only two species of *Andreaea* have been reported for the state.

The order Bryales is represented in the state by at least 123 genera; several of these, however, are either monotypic or known in this region by a single species. About 375 species of the Bryales have been reported; many of them are quite common and widely distributed in both the Lower and Upper Peninsulas, but a few have been found only once in Michigan. The occurrence of many species is determined by habitat. Though latitude and altitude determine to a certain extent the range and distribution of the various species, the most common factors are those associated with the substratum, the average moisture, and the relative amount of light. In every one of the 68 counties of the Lower Peninsula, many combinations of these fundamental conditions will be found. Cool, moist woods afford, perhaps, the most favorable habitats, such as bases of trees, logs, erratic boulders with or without a thin covering of soil, occasional swales, and the banks and even the bottoms of brooks. Some species are essentially limited to open swamps, while a number are found in drier, more open situations, often produced by the cutting of the original timber; other species prefer such artificial habitats as meadows, fallow fields, gravel banks, roadsides, and old quarries.

Intensive collecting in the Douglas Lake region of Cheboygan and Emmet counties over a number of years has shown that the majority of the Bryales are found in the habitats noted above. In the extensive areas of exposed rock, cliffs, or ledges occurring in parts of the Upper Peninsula may be found an interesting group of mosses whose species are limited to outcropping sandstone. Another group grows more commonly in association with limestone, on the surface or in crevices of such rocks or in seepage from them. Such calcareous habitats are rather localized in the Lower Peninsula; in the Upper Peninsula they occur along the south shore of Mackinac County and on Mackinac Island. A number of rare species have been collected only on the sandstone outcrops along Lake Superior in Alger County east and west of Munising. With the exceptions noted above, the physiography and habitat relationships of the six eastern counties of the Upper Peninsula are not particularly different from those found in the Lower Peninsula. The general level of the country is mostly less than 200 feet above the surface of the Great Lakes. West and north of Marquette the surface forms a tableland rising about 1000 feet above the level of Lake Superior, in places rising still higher to form the so-called mountainous regions known in Marquette County as the Huron Mountains and in Ontonagon County as the Porcupine Mountains; similar geologic formations are found in Houghton and Keweenaw counties (including Isle Royale). Outcrops of granitic rocks are common in the region and harbor a number of mosses found nowhere else in the state.

In Michigan the average annual temperature of the interior portions of the Upper Peninsula differs from that of the southern tier of counties of the Lower Peninsula by at least ten degrees. This results not only from a distance of about 600 miles from the southern border of the state to the tip of Keweenaw County in the Upper Peninsula but also from the higher elevation of the northern region. This climatic difference is undoubtedly responsible, at least in part, for the presence of a number of arctic-alpine bryophytes in the more northern counties bordering Lake Superior. The number of these species is not large, but their occurrence in Michigan is significant.

A few species are apparently disjuncts whose natural ranges are far west of the Great Lakes.

THE LIFE HISTORY OF MOSSES

The primary aim of a beginner using a manual on the mosses of any region is to identify the various mosses which may be seen or collected. Before learning something of the considerable terminology applied to the description of mosses, it is desirable to become acquainted with the life history of the individual moss plant.

The minute spherical bodies produced in the capsule of a moss are called *spores;* when dry, these are blown or scattered in all directions from the parent plant. If a spore falls in a favorable situation, it will germinate by sending out a slender filament, the *protonema,* which by continued branching will normally form a green mat on the substratum. Buds arising on the protonema grow rapidly into leafy green shoots, which soon become anchored at the base by a growth of *rhizoids.* The leafy stems form the conspicuous part of the plant. They constitute, with the protonema and later the sex organs, a phase of the life history called the *gametophyte.* The stems may have few or many leaves, and the leaves are generally one cell thick and often closely overlapping.

At the ends of the stems or on lateral shoots, more or less modified leaves enclose clusters of reproductive organs. The male organs (*antheridia*) produce numerous *sperms;* each female organ (*archegonium*) contains one *egg* cell. The antheridium is a stalked, thin-walled structure resembling a cucumber before the discharge of sperms. At maturity it swells by absorbing water, breaks open, and liberates curved, biciliate sperms. The *archegonium* (fig. 1) is a stalked, flask-shaped body with an enlarged base, the *venter,* and a tapering *neck.* The neck contains a row of sterile cells (*canal cells*) above the single egg cell in the venter. When the archegonium is ripe, the canal cells disintegrate, forming a slimy protoplasmic mass, some of which may ooze out at the top of the neck. If nearby surfaces are wet, sperms may be attracted to the opening of the neck and swim down it. If a sperm reaches the egg, it normally unites with it to effect fertilization and form a *zygote,* the first cell of the sporophytic generation. The single-celled *sporophyte* by successive division develops tissue above and below it. Below, an absorbing organ is formed that penetrates the tissue of the gametophyte; above, a slender stalk, the young *seta,* grows up through the neck of the archegonium whose venter, meanwhile, has become considerably enlarged. The upward growth of the seta finally pulls the venter loose from its base and carries it upward, enclosing the young capsule at the tip of the seta. (The young sporophytes at this stage

[9]

are sometimes spoken of as "spears.") When the seta has reached its full height, the enlarged venter capping the spore case is known as the *calyptra,* a thin-walled hood which easily comes off when the capsule is ripe. At the top of the capsule a lid, *operculum,* is formed. Usually one or more rows of hygroscopic teeth are developed underneath the lid, forming the *peristome.* Surrounding a core of thin-walled cells (the *columella*) in the upper part of the maturing capsule, a layer of sporogenous tissue is developed, giving rise to spore mother cells, which by reduction division form the spores. These finally fill most of the *urn* (that part of the capsule between the operculum and the *neck*). After the operculum falls away, the spores may be held back by the teeth until climatic conditions are favorable for their release.

The life cycle may be written briefly as follows (See also fig. 1):

spore → protonema → male plant ⟶ antheridium → sperms ⟍ female plant → archegonium → egg cell ⟋ zygote (first cell of sporophyte) ⟶ mature sporophyte ⟶ capsule ⟶ spore mother cells ⟶ spores. The 2nd, 3rd, 4th and 5th stages constitute the *gametophyte;* the 6th, 7th, 8th and 9th stages, the *sporophyte.*

To Study Mosses

It can be assumed that anyone attempting to use this manual will already have some knowledge of the higher plants, gained either individually or in high school or college. Usually he will have some acquaintance with botanical terminology as it refers to leaves, stems, roots, etc. He should know how to use a glossary of botanical terms. If he has had a course in elementary botany or zoology, he will usually own a dissecting set and should have no difficulty in examining the parts of a moss plant or in making a temporary water mount for examination under the microscope. Those who have had a course in bryology will usually have had sufficient training in methods and techniques to use at once the keys to genera and species presented in this manual. For thorough work, it is necessary to have access to some sort of dissecting microscope, as well as a compound microscope.

To a beginner mosses are apt to look very much alike. For this reason the initial step of making a small collection of mosses, properly dried and preserved, is one of the best approaches to the subject. The habit of taking field notes on approximate size, methods of growth, character of substratum, relative age, locality, and date of collection will form a basic training for the recognition of numerous genera and species in the field; also the beginner will learn to make allowances for differences in the appearance of a moss owing to age or to wet or dry conditions. A hand lens may be used with profit in the field to discover points of interest not visible to the naked eye.

Freshly collected specimens may be wrapped in pieces of newspaper or, better still, placed in separate envelopes, provided that the specimens are not too wet. Specimens should be dried under light pressure in a plant press or between sheets of blotting paper. Clean specimens are very desirable; some dirt is more easily removed after drying than before. After a specimen is dry, it may be enclosed in a packet for future study or for reference. A packet may be made from a rectangular sheet of paper. The lower edge of the sheet is folded part way toward the top, the upper part then turned forward and the ends folded backward. Packets may be glued, pinned, or stapled to herbarium sheets or filed in shoe boxes. Dried specimens are sometimes glued on cards without being packeted. For further information on the preparation of dried specimens, see Conard and committee (1945) and Dore (1953). Directions for preparing the parts of a moss plant for examination with a microscope or for making permanent mounts have been set forth in a number of books and articles. Each worker adopts the technique or combination of methods best suited to his own needs. The following references may be consulted: preparation for examination (Grout, 1947); sectioning of leaves (Hutchinson, 1954); making permanent mounts (Sayre, 1941; Anderson, 1954).

Generic Key

1. Leaf cells of 2 kinds, large, empty, hyaline cells
   enclosed in a meshwork of narrow, green cells . *Sphagnum*, p. 21
1. Leaf cells uniform 2
   2. Small plants in dark, usually black or brown
      tufts on acid rocks; capsule opening by 4
      longitudinal slits . . . . *Andreaea*, p. 33
   2. Plants small to large, mostly green, in
      various habitats; capsules opening by an
      operculum or very rarely indehiscent 3
3. Stems ± erect, not or sparsely branched,
   usually crowded in tufts; sporophytes terminal (Acrocarpi) . . . 4
3. Stems prostrate and matted, freely branched;
   sporophytes lateral (see also *Grimmia, Rhacomitrium,*
   and *Drummondia*) (Pleurocarpi) 76

ACROCARPI

4. Plants minute and inconspicuous; protonema
   mostly ± persistent; capsules mostly
   indehiscent 67
4. Plants minute to large (but rarely less than
   2 cm. high); protonema persistent;
   capsules operculate 5
5. Peristome well developed 6
5. Peristome lacking or poorly developed 60
   6. Leaves 2-ranked; peristome single 7
   6. Leaves several-ranked; peristome
      single or double 8
7. Leaves apparently split at base into 2 laminae which
   clasp the stem and the leaf above . . . . . *Fissidens*, p. 34
7. Leaves subulate from a broader, sheathing base
   (not split into 2 laminae) . . . . *Distichium*, p. 40
   8. Plants grayish or whitish; leaves
      fleshy, consisting mostly of costa . . . *Leucobryum*, p. 58
   8. Plants of various colors, not grayish or
      whitish; leaves generally thin; costa narrow 9
9. Plants blackish- or brownish-green 10
9. Plants green or yellow-green 13
   10. Plants growing on rocks; leaves often ending
      in a hyaline tip or awn; peristome single 11
   10. Plants growing mostly on bark; leaves not
      ending in a hyaline tip or awn; peristome double 12
11. Lower leaf cells linear with thickened, strongly
    sinuous walls; capsules exserted . . . . *Rhacomitrium*, p. 76
11. Lower cells short- or long-rectangular; walls not or
    slightly sinuous; capsules immersed or exserted . *Grimmia*, p. 73
    12. Leaves mostly crisped when dry; leaf cells
       short and pale in 1–2 rows at the basal
       margins, forming an indistinct border;
       capsules exserted . . . . *Ulota*, p. 112
    12. Leaves usually not crisped, not bordered
       at base; capsules immersed to exserted . *Orthotrichum*, p. 108

13. Capsules subglobose 14
13. Capsules more than 1.5:1 18
   14. Plants on bark, with creeping, pinnately
      branched stems; capsules erect and
      symmetric . . . . *Drummondia*, p. 114
   14. Plants on soil or rock, with erect or ascending
      stems, not pinnately branched; capsules asymmetric
      and mostly strongly inclined 15
15. Stems 3–angled; capsules erect or
    somewhat inclined . . . . *Plagiopus*, p. 104
15. Stems not noticeably 3–angled; capsules
    horizontal to subpendent 16
   16. Capsules very small, smooth;
      peristome single . . . . . . *Catoscopium*, p. 103
   16. Capsules larger, furrowed; peristome double 17
17. Leaves lanceolate to ovate; peristome with
    cilia; water-loving mosses . . . . . . *Philonotis*, p. 105
17. Leaves linear to subulate; cilia rudimentary or
    lacking; plants of drier habitats . . . . .*Bartramia*, p. 104
   18. Capsules with a conspicuously enlarged
      hypophysis; peristome teeth often in 2's or 4's 19
   18. Capsules without an hypophysis 20
19. Hypophysis much wider than the urn; peristome teeth
    in 2's at dehiscence . . . . *Splachnum*, p. 82
19. Hypophysis little wider than the urn; peristome
    teeth in 4's at dehiscence . . . . . . *Tetraplodon*, p. 82
   20. Peristome of 4 triangular teeth . . . . *Tetraphis*, p. 34
   20. Peristome of more than 4 teeth 21
21. Peristome of 32–64 blunt teeth joined at their
    tips to a circular membrane 22
21. Peristome teeth 16 or 32, not joined at their
    tips to a membrane 24
   22. Capsules angled . . . . *Polytrichum*, p. 193
   22. Capsules cylindric 23
23. Calyptra hairy; leaves not crisped when dry;
    lamellae on both costa and lamina . . . . *Pogonatum*, p. 192
23. Calyptra not hairy; leaves crisped when dry;
    lamellae on costa only . . . . *Atrichum*, p. 191
   24. Peristome of 32 spirally twisted filaments 25
   24. Peristome of 16 teeth 27
25. Leaves elliptic or obovate, mostly rounded at the
    apex; costa usually excurrent as a hair-point . . *Tortula*, p. 71
25. Leaves narrower, mostly widest at base, usually
    tapered to an acute point, not hair-pointed 26
   26. Leaf margins plane or erect; basal cells
      of leaf abruptly enlarged and thin-walled, forming
      a V-shaped border at the shoulders . . . *Tortella*, p. 63
   26. Leaf margins revolute; basal cells not particularly
      enlarged or thin-walled, merging with
      the upper cells . . . . *Barbula*, p. 66
27. Peristome single (or apparently so; in some
    Encalyptae the endostome is pale and adherent
    to the exostome), rarely lacking 28
27. Peristome double 49

28. Neck of the capsule as long as the
    urn or longer . . . . . .   *Trematodon,* p. 40
28. Neck shorter than the urn 29
29. Calyptra large, entirely covering the cylindric
    capsule, fringed; basal cells of leaf with
    thickened and colored end walls . . . . *Encalypta,* p. 59
29. Calyptra and basal cells otherwise 30
    30. Peristome (consisting of endostome only) with
        16 narrow segments from a low basal membrane;
        fruiting stems short, exceeded by subfloral
        innovations, the sporophytes thus appearing
        lateral . . . . *Mielichhoferia,* p. 84
    30. Peristome otherwise; fruiting stems well
        developed and sporophytes clearly terminal 31
31. Peristome teeth usually entire (lacking in
    Seligeria doniana) 32
31. Peristome teeth ± divided 37
    32. Leaves mostly subulate from a broader base 33
    32. Leaves mostly broader throughout, often
        lanceolate or oblong 34
33. Plants minute; alar cells not enlarged . . . *Seligeria,* p. 44
33. Plants small to medium-sized; alar cells
    inflated, forming auricles . . . . . . *Blindia,* p. 46
    34. Alar cells ± differentiated . . . *Dicranoweisia,* p. 52
    34. Alar cells not differentiated 35
35. Capsules deeply 8-plicate when dry;
    leaf margins plane . . . . . . *Rhabdoweisia,* p. 48
35. Capsules not deeply plicate; leaf
    margins ± revolute 36
    36. Leaves oblong, with margins recurved at the
        middle, serrulate above;
        peristome smooth . . . . . . *Oreoweisia,* p. 49
    36. Leaves lanceolate or oblong from a broader
        base, with margins revolute below, not or sparsely
        toothed at the tip; peristome papillose .   *Didymodon,* p. 68
37. Peristome teeth divided 1/2–3/4 to the base 38
37. Peristome teeth divided nearly to the base 43
    38. Leaves setaceous . . . . . *Dicranodontium,* p. 57
    38. Leaves mostly narrow but not setaceous 39
39. Plants bluish- or glaucous-green . . . . *Saelania,* p. 42
39. Plants not bluish- or glaucous-green 40
    40. Alar cells enlarged and inflated . . . *Dicranum,* p. 52
    40. Alar cells not enlarged and inflated 41
41. Leaves often bistratose at the
    upper margins . . . . . . *Oncophorus,* p. 50
41. Leaves unistratose 42
    42. Leaves broadly oblong; costa ending below the apex;
        upper cells subquadrate, mammillose on both surfaces,
        extending downward in several marginal rows
        nearly to the base . . . . . *Dichodontium,* p. 49
    42. Leaves narrower, ± subulate; costa subpercurrent
        to excurrent; upper cells short or ± elongate,
        smooth; basal marginal cells not short . . *Dicranella,* p. 47

[14]

[15]

PLEUROCARPI

[16]

[18]

[19]

Descriptions

ORDER I. SPHAGNALES

Plants of bogs or swampy margins of lakes and ponds, particularly in cooler regions, grayish-green or yellowish to reddish, sometimes purple-brown or nearly black, continuing apical growth indefinitely and thus forming deep beds of peat. Stems usually erect in dense tufts or large hummocks. Branches in fascicles of 2–6 or rarely more, spirally arranged at regular intervals, crowded at the stem tips; antheridia in the axils of highly colored branch leaves; capsules globose, operculate, borne terminally on an extension of the stem. The order consists of a single genus, *Sphagnum* L.

1. Cortical cells of stems and branches with spiral fibrils 2
1. Cortical cells without fibrils 5
    2. Chlorophyll cells of branch leaves elliptic or oval in cross-section, included or about equally exposed 3
    2. Chlorophyll cells of branch leaves isosceles-triangular in section, with the base exposed on the inner surface . 4. *S. palustre*
3. Chlorophyll cells broadly elliptic in cross-section, with thin side- and end-walls, reaching both surfaces . . 3. *S. papillosum*
3. Chlorophyll cells of branch leaves short-oval in cross-section, or if long-oval, with thick end-walls, included 4
    4. Chlorophyll cells of branch leaves short-oval in cross-section, entirely included . . . . . 1. *S. magellanicum*
    4. Chlorophyll cells long-oval in cross-section, with thick end-walls, thus reaching both surfaces . . . 2. *S. centrale*
5. Cortical cells of branches uniform, each with a pore at the upper end . . . . 5. *S. compactum*
5. Cortical cells of branches of 2 kinds, with larger retort cells with a neck and a pore, or if uniform, without pores 6
    6. Branches in fascicles of 6–12 . . . . . 6. *S. wulfianum*
    6. Branches never more than 6 7
7. Stem leaves with the membrane of hyaline cells mostly resorbed on the outer surface, on the inner with membrane gaps only in apical cells 8
7. Stem leaves with the membrane of hyaline cells usually not greatly resorbed on the outer surface, or if so, resorbed on both surfaces resulting in a lacerate apex 9
    8. Plants large; branch leaves ovate-hastate, mostly squarrose . . . . 7. *S. squarrosum*
    8. Plants smaller; branch leaves ovate-lanceolate, usually imbricate . . . . 8. *S. teres*
9. Chlorophyll cells of branch leaves exposed only or more broadly on the outer surface, or if with central lumen and ± equal exposure, the pigment brown 10
9. Chlorophyll cells of branch leaves exposed only or more broadly on the inner surface, or if with central lumen and ± equal exposure, the pigment red 15

1. *Sphagnum magellanicum* Brid. — Plants large, coarse, deep- to purple-red. Cortical cells of the stem fibrillose, with 1–2 pores. Stem leaves wider at apex than at base, 1.5–2:1, denticulate, with hyaline cells only rarely divided and usually lacking fibrils. Branch leaves broadly ovate with a denticulate border; hyaline cells small and narrow, with a few thin fibrils, only slightly convex on both surfaces and thus square to rectangular in cross-section, on the outer surface with 4–10 large, elliptic pores, on the inner with few corner pores in apical and lateral regions of the leaf, the pores distinct but not ringed; chlorophyll cells small-elliptic in section and entirely

enclosed (or rarely slightly exposed). — Labrador to Alaska, south to the Gulf of Mexico and California; Europe, Asia, and South America.

Sphagnum compactum is usually much shorter and dirty-white to green or brownish-green, with non-fibrillose cortical cells. The chlorophyll cells of *S. centrale* have thickened end-walls, and the plants are usually yellowish-brown.

ILLUSTRATIONS: Figure 3. Jennings, *Mosses of Western Pennsylvania* (ed. 2), Pl. 3. Welch, *Mosses of Indiana*, fig. 10.

Upper Peninsula: Dickinson Co., *Davis*. Gogebic Co., *Steere*. Keweenaw Co., *Allen & Stuntz*. Mackinac Co., *Ehlers*. Marquette Co., *Mains*. Ontonagon Co., *Nichols & Steere*. Schoolcraft Co., *Ehlers*.

Lower Peninsula: Cheboygan Co., *Praeger*. Ingham Co., *Marshall*. Iosco Co., *Cantlon & Gillis*. Mecosta Co., *Schnooberger*. Washtenaw Co., *Steere*.

2. *Sphagnum centrale* C. Jens. — Plants robust, green to bright yellow-brown. Wood cylinder of stem green to brownish; cortical cells in 3–4 layers, their walls thin and reinforced by spiral thickenings, the outer cells long-rectangular to pentagonal, usually with 1 large or 2 smaller pores (or occasionally with 4 small pores). Stem leaves large, broadly spatulate, very concave at apex, bordered by 2–4 rows of short, hyaline cells, denticulate throughout or finely meshed across the broad apex; hyaline cells usually lacking fibrils, rarely divided, 6–8: 1 at base, 1–2: 1 at apex. Branches in fascicles of 4–5, robust and spreading, the others weak, pendent, and appressed to the stem; cortical cells in 1 layer, fibrillose, usually with 1 large pore at the upper end; branch leaves large, concave, and cucullate, loosely appressed, denticulate; hyaline cells 8–10:1 at base, 2:1 at apex, fibrillose, smooth where overlying chlorophyll cells, convex on the outer surface (up to ¼ the diameter), less so on the inner surface, with 2–6 round to oval, ringed pores at the ends and along the commissures on the inner surface, on the outer surface similar, with 2–3 often grouped at contiguous cell corners; chlorophyll cells in section long-elliptic with thick end-walls reaching both surfaces, the lumen broadly oval to narrowly elliptic; resorption furrow present. — Widespread in northern North America; Asia, Europe, and the Azores.

ILLUSTRATION: Figure 4.

Upper Peninsula: Mackinac Co., *Nichols & Ehlers*.
Lower Peninsula: Cheboygan Co., *Nichols*.

3. *Sphagnum papillosum* Lindb. — Plants varying greatly in size, green or deep-brown or nearly black. Cortical cells of stem with 1–2

small, distinct pores; fibrils reduced or nearly lacking. Stem leaves large and long-lingulate or smaller and short-lingulate, the border denticulate, narrowed at apex; hyaline cells often divided. Branch leaves ovate, denticulate; hyaline cells with 4–10 pores per cell on the outer surface (appearing in 3's in the angles of contiguous cells) and on the inner surface with small pores in cell corners (and occasionally large, round pores in the centers in apical and lateral regions), 1–4 per cell, where overlying chlorophyll cells usually densely and finely papillose, strongly convex on the outer surface; chlorophyll cells about equally exposed, lenticular to truncately elliptic in cross-section with a central, elliptic lumen and thick walls. — Labrador to New Jersey and the Great Lakes; Alaska and British Columbia; northern Europe and Asia.

ILLUSTRATIONS: Figure 5. Jennings, *Mosses of Western Pennsylvania* (ed. 2), Pl. 2.

Upper Peninsula: Keweenaw Co., *Cooper*. Schoolcraft Co., *Ehlers*.
Lower Peninsula: Cheboygan Co., *Praeger*. Mecosta and Montcalm counties, *Schnooberger*. Washtenaw Co., *Grassl*.

4. *Sphagnum palustre* L. — Plants usually green (grayish-white when dry) to slightly brownish. Cortical cells of stem with 1–4 pores (occasionally up to 10). Stem leaves large, long-lingulate to spatulate, broadly rounded at apex, with a denticulate to fimbriate border, the hyaline cells not divided. Branch leaves ovate, denticulate; hyaline cells on the outer surface with large pores in the ends and smaller pores along the commissures, on the inner surface with small pores in the angles and a few larger, central pores, somewhat convex on the outer surface, with inner walls smooth; chlorophyll cells narrowly isosceles-triangular in section, with the base exposed on the inner surface (forming a straight line). — Newfoundland to the Great Lakes and the Gulf of Mexico; Europe, Asia, and South America.

ILLUSTRATIONS: Figure 6. Jennings, *Mosses of Western Pennsylvania* (ed. 2), Pl. 1. Welch, *Mosses of Indiana*, fig. 11.

Upper Peninsula, *Osborn*. Delta Co., *Gleason*. Houghton Co., *Mugford*. Keweenaw Co., *Povah*. Mackinac Co., *Ehlers*. Ontonagon Co., *Darlington*.
Lower Peninsula: Berrien Co., *Andrews*. Cheboygan Co., *Wheeler*. Clinton Co., *Janson*. Ingham Co., *Wheeler*. Leelanau Co., *Darlington*. Oakland Co., *Schnooberger*. St. Joseph Co., *Wheeler*. Van Buren Co., *Schnooberger*. Washtenaw Co., *Steere*.

5. *Sphagnum compactum* DC. — Plants usually short and compact. Stem leaves very small, triangular to triangular-lingulate. Cortical cells of the branches of 1 kind (retort). Branch leaves large,

very sparsely denticulate; hyaline cells with pores in 3's at contiguous corners on the inner surface (sometimes with 3 pores in each cell) and large, rounded pores mostly near the commissures on the outer surface, the fibrils often cross-connected so that the pores are usually enclosed in a fibrillose ring on both surfaces; chlorophyll cells small and entirely included. — Greenland; Labrador to the Great Lakes and south to Florida and Alabama; Alaska and British Columbia; Europe and Asia.

ILLUSTRATIONS: Figure 7. Jennings, *Mosses of Western Pennsylvania* (ed. 2), Pl. 3. Welch, *Mosses of Indiana*, fig. 13.

Upper Peninsula: Marquette Co., *Nichols*.
Lower Peninsula: Cheboygan Co., *Praeger*. Roscommon Co., *Cantlon & Gillis*.

6. *Sphagnum wulfianum* Girg. — Plants medium-sized to fairly robust, green or tinged with brown or red. Cortical cells of stem in 2–4 layers, small but distinctly differentiated from the small wood-cylinder cells. Stem leaves small, triangular-lingulate, ± concave, often involute at the apex; apical hyaline cells resorbed on both surfaces producing a ± lacerate border across the tip. Branches in fascicles of 6–12, usually 3–6 strongly spreading giving the plants a characteristic appearance; retort cells well developed, with small to large pores. Branch leaves narrowed to the base, long-involute to the toothed apex, reflexed at the tips when dry; hyaline cells quite long just above the leaf base, shorter at the apex, on the outer surface with small, distinct pores along the commissures, on the inner surface with small end and corner pores only, somewhat convex, more so on the outer surface; chlorophyll cells elliptic to oval with ± equal exposure or somewhat greater exposure on the outer surface; resorption furrow none. — Greenland; southeastern Canada and the northern United States to Minnesota; British Columbia; Europe.

ILLUSTRATIONS: Figure 8. Jennings, *Mosses of Western Pennsylvania* (ed. 2), Pl. 2.

Upper Peninsula: Mackinac and Schoolcraft counties, *Ehlers*.
Lower Peninsula: Cheboygan Co., *Praeger*. Montmorency, Oscoda, and Roscommon counties, *Schnooberger*.

7. *Sphagnum squarrosum* Crome — Plants fairly robust, green, whitish-green or yellowish. Stem leaves large, spatulate, lacerate across the wide apex; hyaline cells of apex longer than broad. Branch retort cells not conspicuous. Branch leaves ovate-hastate, noticeably squarrose (or rarely imbricate), not easily flattened out under a cover slip; hyaline cells with pores small in proportion to

cell size, often minutely papillose where overlying the chlorophyll cells which in cross-section are small and triangular to trapezoidal with broader exposure on the outer surface; resorption furrow lacking. — Greenland; Labrador to Pennsylvania and the Great Lakes; Alaska to California; Europe and Asia.

Similar to *S. teres* but generally larger and coarser with branch leaves and their hyaline cells also larger.

ILLUSTRATIONS: Figure 9. Jennings, *Mosses of Western Pennsylvania* (ed. 2), Pl. 70.

Upper Peninsula: Chippewa Co., *Steere*. Delta Co., *Wheeler*. Houghton Co., *Parmelee*. Keweenaw Co., *Povah*. Marquette Co., *Barlow*. Ontonagon Co., *Nichols & Steere*.

Lower Peninsula: Alpena Co., *Wheeler*. Cheboygan and Emmet counties, *Praeger*. Leelanau Co., *Darlington*. Macomb Co., *Cooley*. Washtenaw Co., *Schnooberger*.

8. *Sphagnum teres* (Schimp.) Ångstr. — Stem leaves similar to those of *S. squarrosum* but with hyaline cells at the apex broad and short and the margin narrow throughout. Branch leaves ovate-lanceolate, usually imbricate (rarely somewhat squarrose), easily flattened out under a cover slip; hyaline cells with pores few and large in proportion to cell size, sometimes papillose where overlying the chlorophyll cells which are small and triangular to trapezoidal with broader exposure on the outer surface; resorption furrow none. — Greenland; Labrador to New Jersey and Michigan; Alaska to California and Colorado; Europe.

ILLUSTRATIONS: Figure 10. Jennings, *Mosses of Western Pennsylvania* (ed. 2), Pl. 3.

Upper Peninsula: Keweenaw Co., *Cooper*.
Lower Peninsula: Cheboygan Co., *Praeger*. Washtenaw Co., *Steere*.

9. *Sphagnum lindbergii* Schimp. — Cortical cells of stem large, thin-walled. Stem leaves usually wider than long, spatulate, concave, lacerate across the broad apex. Retort cells usually single, with inconspicuous necks. Branch leaves often appearing 5-ranked, their hyaline cells long and narrow with pores small, generally restricted to the cell ends, convex, more so on the inner surface; chlorophyll cells triangular with the base exposed on the outer surface or rarely trapezoidal to short-elliptic and exposed on both surfaces; resorption furrow none. — Greenland; Labrador to New York and the Great Lakes; Alaska and British Columbia; Europe.

Sphagnum fimbriatum has chlorophyll cells trapezoidal and exposed on the inner surface and hyaline cells bulging up to 1/2 their diameters on the outer surface; the stem leaves are broader than

[26]

long but lacerate at the sides as well as the apex. *S. girgensohnii* has chlorophyll cells trapezoidal to triangular with inner exposure and a triangular lumen and chlorophyll cells scarcely convex on the inner surface but bulging 1/5–1/3 their diameters on the outer surface; the stem leaves are lingulate, about as long as broad, and lacerate across a slightly narrowed apex.

ILLUSTRATION: Figure 11.

A single specimen is known from Michigan, but the county is unrecorded (in herb. *Schnooberger*).

10. *Sphagnum recurvum* P.–Beauv. — Plants usually bright-green or slightly tinged with brown. Cortical cells of stems scarcely differentiated. Stem leaves small, usually lacking fibrils. Branch leaves with recurved tips and a few undulations when dry; hyaline cells with a few small pores on the inner surface and a few larger pores on the outer surface; chlorophyll cells broadly triangular with the base exposed on the outer surface; resorption furrow none. — Labrador to Michigan and south to Florida and Louisiana; Alaska to Washington and Colorado; Europe, Asia, and South America.

ILLUSTRATIONS: Figure 12A. Jennings, *Mosses of Western Pennsylvania* (ed. 2), Pl. 3. Welch, *Mosses of Indiana*, fig. 14.

Upper Peninsula: Chippewa and Keweenaw counties, *Povah*. Schoolcraft Co., *Ehlers*.
Lower Peninsula: Cheboygan and Emmet counties, *Nichols*. Washtenaw Co., *Steere*.

10a. *Sphagnum recurvum* var. *tenue* Klinggr. — Plants smaller. Branch leaves imbricate, not undulate. — Greenland; Newfoundland to Pennsylvania and across the continent; Europe and Asia.

ILLUSTRATION: Figure 12B.

Upper Peninsula: Chippewa Co., *Thorpe*. Keweenaw Co., *Cooper*.
Lower Peninsula: Cheboygan Co., *Praeger*.

11. *Sphagnum pulchrum* (Lindb.) Warnst. — Plants usually robust, brownish to purplish. Cortical cells of stems differentiated; wood cylinder usually brown. Stem leaves small, triangular, usually quite concave, involute at apex; apical hyaline cells sometimes fibrillose, not divided. Retort cells with quite inconspicuous necks. Branch leaves broad, often ± secund and undulate, involute at apex; hyaline cells with small, ringed pores on the outer surface, usually quite convex on the inner surface, scarcely so on the outer; chloro-

[27]

phyll cells triangular with broader exposure on the outer surface; resorption furrow none. — Labrador to New Jersey and the Great Lakes; northern Europe.

ILLUSTRATION: Figure 13.

Lower Peninsula: Emmet Co. (Big Stone Bay), *Nichols & Ehlers.*

12. *Sphagnum cuspidatum* Ehrh. — Plants delicate to stout, often plumose, growing submerged or in very wet places, pale, yellow or greenish. Cortical cells of stems differentiated, usually in 2 layers. Stem leaves small, concave, triangular (sometimes narrowly so), toothed at apex; apical cells usually somewhat fibrillose. Pendent branches poorly developed and inconspicuous. Branch leaves varying greatly in size but long and narrow, involute to a 4–5-toothed apex; hyaline cells long and narrow, with a few end pores only in the apical region, scarcely convex on the outer surface, slightly so on the inner; chlorophyll cells trapezoidal with broader exposure on the outer surface, the lumen triangular; resorption furrow none. — Newfoundland to Florida and the Great Lakes; Europe and Asia.

ILLUSTRATIONS: Figure 14. Jennings, *Mosses of Western Pennsylvania* (ed. 2), Pl. 4.

Upper Peninsula: Houghton Co., *Mugford.* Schoolcraft Co., *Ehlers.*
Lower Peninsula: Cheboygan Co., *Praeger.* Emmet Co., *Nichols & Ehlers.*

13. *Sphagnum dusenii* C. Jens. — Plants usually robust, yellowish-brown or brown. Cortical cells of stems small, with medium-thick walls. Stem leaves medium-sized, triangular, concave, involute to a blunt apex. Retort cells very long (often 1 above the other), with inconspicuous necks. Branch leaves large, rarely slightly undulate or strongly involute; hyaline cells quite long, with pores usually lacking on the inner surface and several large pores down the middle and usually a few small pores at the corners of cells on the outer surface, slightly convex, more so on the inner surface; chlorophyll cells trapezoidal in section, thin-walled, with the broader base exposed on the outer surface. — Labrador to New York and Wisconsin; Alaska and British Columbia.

ILLUSTRATION: Figure 15.

Upper Peninsula: Dickinson and Houghton counties, *Davis.* Keweenaw Co., *Hermann.*

14. *Sphagnum subsecundum* Nees — Plants exceedingly variable in all characteristics. Stem leaves usually rather small, narrowly lingulate, and involute to a blunt apex, but sometimes ovate and

concave or very similar to branch leaves. Branch leaves slightly secund to imbricate, ovate and strongly concave or narrower and deeply involute, toothed at the apex; hyaline cells usually with many, small, ± round pores along the commissures on the outer surface (but sometimes much reduced) and normally only a few, small, round pores on the inner surface, about equally convex or more so on the inner surface; chlorophyll cells elliptic and equally exposed or triangular to trapezoidal with broader exposure on the outer surface (or nearly equal); resorption furrow none. — Greenland; Labrador to the Great Lakes and the Gulf of Mexico; Washington to California; Mexico; Europe and Asia.

ILLUSTRATION: Figure 16.

Upper Peninsula: Alger Co., *Bailey*. Houghton Co., *Mugford*. Keweenaw Co., *Povah*. Schoolcraft Co., *Ehlers*.

Lower Peninsula: Cheboygan Co., *Nichols & Ehlers*. Roscommon Co., *Cantlon & Gillis*.

15. *Sphagnum fimbriatum* Wils. — Plants slender, green or yellowish. Cortical cells of stem usually with 1 large pore at the upper end. Stem leaves broader than long, concave, lacerate across the broad apex and down the sides; hyaline cells rhomboidal, often divided (sometimes repeatedly) in side and apical regions. Retort cells long and rather narrow, with inconspicuous necks. Branch leaves medium-sized, ovate, strongly involute to a narrow, toothed apex; hyaline cells with large pores on the inner surface and more numerous, smaller, elliptic pores on the outer surface, convex, slightly more so on the outer surface; chlorophyll cells trapezoidal with broader exposure on the inner surface; resorption furrow none. — Greenland; Labrador to Alaska and south to Maryland and California; Europe, Asia, and South America.

ILLUSTRATIONS: Figure 17. Jennings, *Mosses of Western Pennsylvania* (ed. 2), Pl. 4. Welch, *Mosses of Indiana*, fig. 17.

Upper Peninsula: Alger Co., *Gleason & Steere*.
Lower Peninsula: Washtenaw Co., *Schnooberger*.

16. *Sphagnum girgensohnii* Russ. — Plants usually fairly robust, green, yellowish, or brownish. Cortical cells of stems medium-large, thin-walled, usually porose. Stem leaves medium-sized, lingulate, as long as broad or slightly longer, lacerate across the slightly narrowed apex. Retort cells usually single, with conspicuous necks. Branch leaves rather small, ovate, involute at the toothed apex; hyaline cells with numerous elliptic pores along the commissures in side and apical regions on the outer surface and 3–5 large, round

pores on the inner surface, very convex on the outer surface; chloro-phyll cells trapezoidal with broader exposure on the inner surface; resorption furrow none. — Greenland; Labrador to North Carolina and Minnesota; Alaska to Oregon: Europe and Asia.

ILLUSTRATIONS: Figure 18. Jennings, *Mosses of Western Pennsylvania* (ed. 2), Pl. 7. Welch, *Mosses of Indiana*, fig. 18.

Upper Peninsula: Chippewa Co., *McVaugh*. Keweenaw Co., *Allen & Stuntz*. Mackinac Co., *Ehlers*. Marquette Co., *Nichols*. Ontonagon Co., *Nichols & Steere*.
Lower Peninsula: Cheboygan Co., *Praeger*. Leelanau Co., *Darlington*. Washtenaw Co., *Steere*.

17. *Sphagnum robustum* (Russ.) Röll — Plants variable, usually quite robust, green or red-tinged. Cortical cells of stems large, thin-walled, with 1 pore at the upper end (most easily observed on staining). Stem leaves large, lingulate-ovate, longer than broad, usually slightly lacerate at the blunt apex; border at base usually reddish; hyaline cells sometimes divided. Retort cells with conspicuous necks. Branch leaves small, involute to a narrow, toothed apex; hyaline cells usually with 4–8 elliptic pores on the outer surface; chlorophyll cells triangular in section with exposure on the inner surface; resorption furrow none. — Greenland; Labrador to New York and across the continent; Europe.

ILLUSTRATION: Figure 19.

Upper Peninsula: Keweenaw Co., *Allen & Stuntz*. Mackinac Co., *Ehlers*.
Lower Peninsula: Cheboygan Co., *Praeger*.

18. *Sphagnum fuscum* (Schimp.) Klinggr. — Plants delicate or rarely robust, in compact, usually brownish tufts. Cortical cells of stems medium-sized, thin-walled. Stem leaves medium-sized, broadly lingulate, 1.5–2:1; many hyaline cells divided. Retort cells long, with long necks. Branch leaves small, involute at the toothed apex; hyaline cells with 3–8 elliptic pores near the commissures on the outer surface and 2–4 in the ends or side corners on the inner surface, strongly convex on the outer surface only; chlorophyll cells triangular to trapezoidal with a small, triangular lumen, exposed on the inner surface; resorption furrow none. — Greenland; Labrador to New York and across the continent; Europe.

ILLUSTRATIONS: Figure 20. Jennings, *Mosses of Western Pennsylvania* (ed. 2), Pl. 8.

Upper Peninsula: Keweenaw Co., *Allen & Stuntz*. Mackinac Co., *Ehlers*. Marquette Co., *Nichols*.
Lower Peninsula: Cheboygan Co., *Praeger*.

19. *Sphagnum warnstorfianum* DuRietz — Plants delicate, green or reddish. Cortical cells of stems medium-sized, thin-walled. Stem leaves medium-sized, lingulate-ovate, only slightly concave, with many hyaline cells divided. Retort cells usually single and well developed, with long necks ending in a small pore. Branch leaves appearing 5-ranked, small, narrow, involute to a rather broad, toothed apex; hyaline cells with 2–4 very small, ringed pores per cell on the outer surface and 2–6 in the ends and side corners on the inner surface, more strongly convex on the outer surface; chlorophyll cells triangular to trapezoidal, with broader exposure on the inner surface; resorption furrow none. — Greenland; Labrador to New York and across the continent; Florida; Europe.

ILLUSTRATIONS: Figure 21. Jennings, *Mosses of Western Pennsylvania* (ed. 2), Pl. 6 (as *S. capillaceum*).

Upper Peninsula: Keweenaw Co., *Allen & Stuntz*. Mackinac Co., *Ehlers*.
Lower Peninsula: Cheboygan Co., *Praeger*.

20. *Sphagnum capillaceum* (Weiss) Schrank — Plants generally delicate, in green or more often reddish, compact tufts. Cortical cells of stems large, thin-walled. Stem leaves large, about 2:1, ovate, quite concave, involute at the apex, with the border usually continuing across the apex. Retort cells good-sized, with conspicuous necks. Branch leaves narrowly ovate, involute at the toothed apex; hyaline cells with 5–12 elliptic pores along the commissures on the outer surface and on the inner surface with small end pores and 2–5 larger pores in the lower side regions, strongly convex on the outer surface only; chlorophyll cells triangular to narrowly trapezoidal, with broader exposure on the inner surface; resorption furrow none. — Greenland; Labrador to Virginia and across the continent; Europe, Asia, and South America.

ILLUSTRATIONS: Figure 22A. Jennings, *Mosses of Western Pennsylvania* (ed. 2), Pl. 5 (as *S. acutifolium* var. *viride*).

Upper Peninsula: Alger Co., *Wheeler*. Houghton Co., *Mugford*. Keweenaw Co., *Allen & Stuntz*. Mackinac Co., *Ehlers*. Marquette Co., *Nichols*. Schoolcraft Co., *Ehlers*.
Lower Peninsula: Cheboygan Co., *Praeger*. Ingham Co., *Marshall*. Iosco Co., *Cantlon*. Roscommon Co., *Mains*. Washtenaw Co., *Schnooberger*.

20a. *Sphagnum capillaceum* var. *tenellum* (Schimp.) Andr. — Delicate plants in dense tufts, usually more deeply red than the species. Stem leaves lingulate, with slight laceration or a single, short rent at the apex; hyaline cells nearly all divided. Branch leaves often slightly secund, generally shorter, with hyaline cells larger than in the species. — Ranging with the species, south to South Carolina.

ILLUSTRATION: Figure 22B.

Upper Peninsula: Mackinac Co., *Ehlers.*
Lower Peninsula: Cheboygan Co., *Praeger.* Washtenaw Co., *Steere.*

21. *Sphagnum plumulosum* Röll — Plants generally quite robust, usually green or yellow, sometimes tinged with brown or purple. Cortical cells of stems medium-sized, thin-walled. Stem leaves medium-sized, long-triangular, about 2:1, involute at the blunt, toothed apex. Retort cells large, with large pores. Branch leaves medium-sized to large, long-ovate, concave, involute at the broad, toothed apex, sometimes slightly squarrose; hyaline cells with 3–10 large, oval pores on the outer surface and 1–4 small pores on the inner surface, more strongly convex on the outer surface; chlorophyll cells triangular to narrowly trapezoidal with broader inner exposure; resorption furrow none. — Greenland; Labrador to New Jersey and Michigan; British Columbia and California; Europe and Asia.

ILLUSTRATIONS: Figure 23. Jennings, *Mosses of Western Pennsylvania* (ed. 2), Pl. 5.

Upper Peninsula: Keweenaw Co., *Allen & Stuntz.*
Lower Peninsula: Cheboygan Co., *Nichols.* Emmet Co., *Nichols & Ehlers.* Roscommon Co., *Cantlon & Gillis.*

22. *Sphagnum tenerum* Sull. & Lesq. — Plants usually robust, in compact, yellowish or pink-tinged tufts. Cortical cells of stems medium-sized, thin-walled. Stem leaves medium to large, triangular, ovate, slightly concave, involute at the narrow, lacerate apex; hyaline cells 6–8:1 at apex, often weakly fibrillose, divided, porose on both surfaces. Retort cells with conspicuous necks and small pores. Branch leaves large, ovate-lanceolate, involute and toothed at the apex; border entire; hyaline cells large, with 5–15 narrowly elliptic pores along the commissures on the outer surface and on the inner surface with small end pores in the apical region and 7–8 large, round pores in the lower side regions, convex on both surfaces, strongly so on the outer; chlorophyll cells very small, narrowly trapezoidal to triangular with the base exposed on the inner surface; resorption furrow none. — Newfoundland to Alabama and Michigan.

ILLUSTRATION: Figure 24.

Upper Peninsula: Mackinac Co., *Ehlers.*
Lower Peninsula: Cheboygan Co., *Nichols.* Washtenaw Co., *Steere.*

[32]

ORDER II. ANDREAEALES

Small, dark-green to red-brown or black mosses growing in tufts on non-calcareous rocks; stems slender, forked, brittle when dry; leaves small, often papillose. Capsule ovoid-oblong, little exserted on an extension of the stem tip serving as a seta, dehiscing by means of 4 longitudinal slits; operculum and peristome none. Represented in North America by a single genus, *Andreaea* Hedw.

Leaves ecostate; spores 20–30 μ . . . . 1. *A. rupestris*
Leaves costate; spores 30–40 μ . . . . 2. *A. rothii*

1. *Andreaea rupestris* Hedw. — Leaves ovate to lanceolate, ecostate. — Forming dark cushions on exposed rocks, ledges, or cliffs. Across Canada and Alaska south to North Carolina, Colorado and Oregon; Greenland; Europe, Japan, Tasmania, and South America.

ILLUSTRATIONS: Conard, *How to Know the Mosses* (ed. 2), fig. 27a–d. Grout, *Mosses with Hand-Lens and Microscope*, Pl. 6.

Upper Peninsula: Keweenaw Co., *Povah*. Marquette Co., *Nichols*. Ontonagon Co., *Nichols & Steere*.

2. *Andreaea rothii* Web. & Mohr — Leaves lance-acuminate, costate. — On rocks and cliffs. Widespread in eastern North America; British Columbia; Europe.

ILLUSTRATIONS: Figure 25. Grout, *Mosses with Hand-Lens and Microscope*, fig. 11. Jennings, *Mosses of Western Pennsylvania* (ed. 2), Pl. 60.

Upper Peninsula: Marquette Co. (Huron Mts.), *Nichols*.

ORDER III. BRYALES

Capsule terminating a usually well-developed seta, surrounded at the base by a cluster of bracts (forming a perichaetium), usually releasing spores by the separation of a lid (or operculum), generally with 1 or 2 circles of teeth at the mouth (the peristome). The Bryales include nearly all the true mosses. They may be separated into the ACROCARPI, whose sparsely branched stems grow erect, usually in tufts, and have the sporophyte terminal on a stem or ordinary branch, and the PLEUROCARPI, which grow ± prostrate, usually in interwoven mats, with stems freely branched and sporophytes lateral on a special short branch.

TETRAPHIDACEAE

Slender, erect, ± tufted plants; leaves lanceolate to ovate, with rounded-quadrate cells; costa single, reaching nearly to the leaf apex or faint; setae elongate; capsules erect and symmetric; peristome consisting of 4 teeth; calyptra mitrate. Represented in our area by 2 species of *Tetraphis* Hedw.

Plants up to 3 cm. tall, usually on very rotten wood; sterile
   plants usually bearing gemmae in apical cups;
   costa strong . . . . 1. *T. pellucida*
Plants very small (less than 5 mm. high), growing in deeply
   shaded crevices of cliffs or banks; sterile stems not
   gemmiferous; costa faint . . . . 2. *T. browniana*

1. *Tetraphis pellucida* Hedw. — Tufts green or brownish, reddish below; gemmae common in cups at the tips of sterile stems; costa well developed; spores in early summer. — Common on very rotten logs and stumps, especially in deep shade. Alaska to California and eastern North America; Europe and Asia.

ILLUSTRATIONS: Figure 26. Conard, *How to Know the Mosses* (ed. 2), fig. 51. Grout, *Mosses with Hand-Lens and Microscope*, fig. 13. Jennings, *Mosses of Western Pennsylvania* (ed. 2), Pl. 30. Welch, *Mosses of Indiana*, fig. 20.

Upper Peninsula: Chippewa Co., *Steere*. Delta Co., *Parmelee*. Gogebic Co., *Bessey*. Houghton Co., *Parmelee*. Keweenaw Co., *Allen & Stuntz*. Mackinac Co., *Parmelee*. Marquette Co., *Nichols*. Ontonagon Co., *Bessey*.

Lower Peninsula: Barry Co., *Parmelee*. Cheboygan Co., *Nichols*. Clinton Co., *Parmelee*. Eaton and Emmet counties, *Wheeler*. Ingham and Leelanau counties, *Darlington*. Macomb Co., *Cooley*. Oakland Co., *Darlington*.

2. *Tetraphis browniana* (Dicks.) Grev. — Plants very small (less than 5 mm. high), scattered or gregarious; leaves narrow; costa faint. — In deeply shaded crevices of calcareous sandstone or in sheltered niches of banks; apparently rare and local. Northeastern United States and adjacent provinces of Canada; Europe.

ILLUSTRATION: Grout, *Moss Flora of North America*, vol. 1, Pl. 3.

Upper Peninsula: Alger Co. (Tannery Falls), *Steere*.

FISSIDENTACEAE

Plants erect, mostly small, green, mostly on soil; stems radiculose at base; leaves distichous, the lower portion apparently split so as to clasp the stem and the base of the leaf immediately above; costa usually well developed, single; cells small, rounded-hexagonal; setae terminal or occasionally lateral; capsules smooth, erect and symmetric or curved and asymmetric; peristome single.

FISSIDENS Hedw.

Capsules mostly exserted; operculum short- to long-rostrate; peristome teeth 16, incurved, forked, papillose; annulus sometimes present; spores mostly small.

1. *Fissidens minutulus* Sull. — Plants very small (1–3 mm. high); upper leaves narrowly lanceolate; border ending well below the apex; costa percurrent; cells rounded-hexagonal, thickened; capsules usually erect and symmetric, 0.7 mm. long; spores in August and September. — On shaded rocks, often in stream beds. Eastern Canada and the United States south to the Gulf of Mexico; Europe.

ILLUSTRATIONS: Grout, *Mosses with Hand-Lens and Microscope*, fig. 23. Jennings, *Mosses of Western Pennsylvania* (ed. 2), Pl. 14. Welch, *Mosses of Indiana*, fig. 32.

Lower Peninsula: Alpena Co., *Robinson & Wells*. Eaton Co., *Parmelee*. Ingham Co., *Darlington*. Washtenaw Co., *Steere*.

2. *Fissidens viridulus* (Web. & Mohr) Wahl. — Plants larger than *F. minutulus* and often difficult to distinguish from *F. bryoides;* upper leaves rounded-obtuse with the border ending somewhat below the apex. — On moist soil or rocks. New England across southern Canada, south to the Middle Atlantic States; Europe, Asia, North Africa, and Macaronesia.

ILLUSTRATION: Grout, *Moss Flora of North America*, vol. 1, Pl. 5c.

Upper Peninsula: Mackinac Co., *Nichols*. Ontonagon Co., *Nichols & Steere*.

3. *Fissidens bryoides* Hedw. — Stems 2–25 mm. high; leaves 1.5–2 mm. long; border confluent with the costa at apex; cells variable in size and shape, 8–12 μ; capsules typically erect and symmetric; spores in winter. — On moist soil, humus, and rotten wood. Canada and the United States, mostly east of the Rocky Mts.; Europe, Asia, and Macaronesia.

ILLUSTRATIONS: Grout, *Mosses with Hand-Lens and Microscope*, fig. 21. Jennings, *Mosses of Western Pennsylvania* (ed. 2), Pl. 14. Welch, *Mosses of Indiana*, fig. 33.

Upper Peninsula: Ontonagon Co., *Nichols & Steere*.
Lower Peninsula: Emmet Co., *Steere*. Ingham Co., *Parmelee*. Washtenaw Co., *Steere*.

4. *Fissidens obtusifolius* Wils. — Small, pale-green plants (2–3 mm. high); upper leaves about 1. mm. long, oblong-lingulate, rounded-obtuse, entire; costa ending a little below the apex; cells pellucid, 7–10 μ; capsule oblong-obovate, erect and symmetric; spores 18–25 μ, mature in autumn. — On wet calcareous rocks. New England and Ontario to Minnesota, south to the Gulf of Mexico; Colorado and Arizona.

ILLUSTRATIONS: Conard, *How to Know the Mosses* (ed. 2), fig. 36a–b. Grout, *Mosses with Hand-Lens and Microscope*, fig. 24. Jennings, *Mosses of Western Pennsylvania* (ed. 2), Pl. 14. Welch, *Mosses of Indiana*, fig. 35.

Upper Peninsula: Luce Co., *Nichols*.
Lower Peninsula: Eaton Co., *Steere*. Huron Co., *Nichols*.

5. *Fissidens taxifolius* Hedw. — Small, light-green plants about 5–20 mm. high; leaves about 2 mm. long, rounded-obtuse and apiculate; costa shortly excurrent; cells 7–10 μ, bulging; setae flexuous; capsules inclined to subpendulous, about 1.5 mm. long; spores 15–20 μ, mature in winter. — On moist clayey soil. Eastern United States and Canada, west to Missouri and south to Florida; Europe, Asia, Macaronesia.

ILLUSTRATIONS: Conard, *How to Know the Mosses* (ed. 2), fig. 39a–b. Grout, *Mosses with Hand-Lens and Microscope,* fig. 26. Jennings, *Mosses of Western Pennsylvania* (ed. 2), Pl. 15. Welch, *Mosses of Indiana,* fig. 38.

Lower Peninsula: Branch Co., *Mains.* Washtenaw Co., *Steere.*

6. *Fissidens osmundioides* Hedw. — Plants dark-green, usually in dense tufts up to 5 cm. high; leaves rounded-obtuse and often apiculate, not bordered; costa ending somewhat below the apex; upper cells 12–20 μ; seta 5–10 mm. long; spores mature from summer to autumn. — On moist, shaded soil and wet rocks. Greenland; northern Canada to Tennessee and Georgia; Europe and Asia.

ILLUSTRATIONS: Grout, *Mosses with Hand-Lens and Microscope,* fig. 25. Welch, *Mosses of Indiana,* fig. 40.

Upper Peninsula: Chippewa Co., *Steere.* Keweenaw Co., *Povah.* Marquette Co., *Nichols.* Ontonagon Co., *Steere.*
Lower Peninsula: Alpena Co., *Robinson & Wells.* Cheboygan Co., *Nichols.* Eaton Co., *Parmelee.* Ingham and Leelanau counties, *Darlington.* Montmorency Co., *Cantlon.* Oakland Co., *Darlington.*

7. *Fissidens subbasilaris* Hedw. — Plants about 5–10 mm. high, with 8–10 pairs of leaves; leaves oblong, obtuse or subacute, apiculate; costa strong, obscured above by short, green, mammillose cells; setae 3–5 mm. long; capsules oblong-cylindric, erect and symmetric; spores 16–18 μ, mature in late autumn. — On earth and bark at base of trees or stumps. Eastern United States south to the Gulf of Mexico.

Lower Peninsula: Kalamazoo Co., *Becker.* Washtenaw Co., *Steere.*

8. *Fissidens adianthoides* Hedw. — Plants relatively large; leaves unevenly serrate, ± bordered by short, pale cells; cells 15 μ or larger, not or slightly bulging; spores in late autumn. — Common in wet woods on soil, logs, and tree bases. Southern Canada across the continent, south to California in the West, Florida in the East; Europe, North Africa, and Macaronesia.

ILLUSTRATIONS: Grout, *Mosses with Hand-Lens and Microscope,* Pl. 9. Welch, *Mosses of Indiana,* fig. 42.

Upper Peninsula: Gogebic Co., *Bessey.* Keweenaw Co., *Povah.* Ontonagon Co., *Darlington & Bessey.*
Lower Peninsula: Cheboygan Co., *Nichols.* Leelanau Co., *Darlington.* Washtenaw Co., *Steere.*

9. *Fissidens cristatus* Wils. — Tufts dark-green, 1–3 cm. high; leaves oblong-lingulate, acute, irregularly serrate, bordered by short, pale cells; costa subpercurrent; cells irregularly hexagonal, bulging,

often bistratose here and there, 6–10 μ, setae 1–4 mm. long; capsules inclined to horizontal, about 2 mm. long; operculum long-rostrate; spores 10–15 μ, mature in late autumn. — On various substrata, frequent in wet woods. Eastern Canada and the United States south to the Gulf of Mexico, west to the Rocky Mts.; Europe, Asia, Azores.

ILLUSTRATIONS: Figure 27. Conard, *How to Know the Mosses* (ed. 2), fig. 38a–d. Jennings, *Mosses of Western Pennsylvania* (ed. 2), Pl. 15. Welch, *Mosses of Indiana*, fig. 43.

Upper Peninsula: Keweenaw Co., *Allen & Stuntz*. Mackinac Co., *Kauffman*. Lower Peninsula: Alpena Co., *Robinson & Wells*. Charlevoix Co., *Parmelee*. Cheboygan Co., *Nichols*. Leelanau Co., *Darlington*. Macomb Co., *Cooley*.

10. *Fissidens grandifrons* Brid. — Plants robust (up to 5 cm. high), rigid, dark-green or black, much branched, often lime-encrusted; leaves linear-lanceolate, stiff, opaque, entire and un-bordered, 2–3 mm. long; costa strong, vanishing in the apex; cells unistratose at the margins, pluristratose within, 7–12 μ, setae up to 1.5 cm. long; capsules nearly symmetric; spores 15–24 μ. — On calcareous rock, submerged. Across Canada and the United States; Mexico and Guatemala; Europe and Asia.

ILLUSTRATIONS: Grout, *Mosses with Hand-Lens and Microscope*, fig. 28. Conard, *How to Know the Mosses* (ed. 2), fig. 35a–b.

Upper Peninsula: Alger Co., *Wheeler*. Mackinac Co., *Nichols*. Lower Peninsula: Charlevoix Co., *Hill*. Cheboygan Co., *Nichols*. Manistee Co., *Hill*. Washtenaw Co., *Steere*.

11. *Fissidens debilis* Schwaegr. — Plants slender, flaccid, green above, blackish-green below; leaves remote, linear-lanceolate, 3–6 mm. long; costa ending below the apex; cells irregularly hexagonal, unistratose; setae shorter than the capsules; capsules 0.5 mm. long; operculum conic-rostrate; spores 18–21 μ, mature in summer. — In limey water, attached to various substrata. New England and Ontario to Florida and Washington to California and Arizona; Europe and Africa.

ILLUSTRATIONS: Conard, *How to Know the Mosses* (ed. 2), fig. 39c. Grout, *Mosses with Hand-Lens and Microscope*, Pl. 10. Welch, *Mosses of Indiana*, fig. 45.

Lower Peninsula: Gratiot Co., *Schnooberger*. Washtenaw Co., *Steere*.

DITRICHACEAE

Plants minute to fairly robust, erect. Leaves mostly lanceolate or subulate, sometimes sheathing at base; costa well developed, sub-percurrent to long-excurrent; cells mostly smooth, not differentiated

[38]

at the basal angles; capsules usually long-exserted, erect or inclined, sometimes asymmetric, mostly operculate; peristome mostly present, the teeth usually variously cleft or perforate.

1. Plants mostly less than 8 mm. high; capsules subglobose
   or pear-shaped, irregularly dehiscent 2
1. Plants generally more than 8 mm. high; capsules oval
   to cylindric, operculate 3
   2. Capsules ovoid, immersed . . . . 1. *Pleuridium*
   2. Capsules pear-shaped, shortly exserted . . . . 2. *Bruchia*
3. Neck of capsule as long as the urn or longer . . . 3. *Trematodon*
3. Neck of capsule much shorter than the urn 4
   4. Plants bluish or glaucous-green . . . . . . 6. *Saelania*
   4. Plants not bluish or glaucous 5
5. Leaves distinctly 2-ranked . . . . 4. *Distichium*
5. Leaves more than 2-ranked 6
   6. Capsules strongly inclined and asymmetric, deeply
      furrowed, somewhat strumose . . . . . . 5. *Ceratodon*
   6. Capsules erect, symmetric or nearly so, not or
      slightly furrowed, not strumose 7
7. Leaves very rough above . . . . 7. *Trichodon*
7. Leaves not or only slightly roughened . . . . 8. *Ditrichum*

1. *Pleuridium* Brid.

Very small, gregarious or cæspitose, yellow-green plants up to 5 mm. high, growing on sandy or clayey soil in open places; upper leaves subulate from an ovate or lanceolate base; capsules ovoid to subglobose, immersed; operculum none; calyptra mostly cucullate.

Pleuridium subulatum (Hedw.) Rabh. — Plants not more than 3 mm. high, in light-green cushions; leaves subulate from a broader base; capsules ovoid; spores maturing in late spring or early summer. — On soil in sandy or clayey fields. Eastern North America; Europe, Asia, North Africa, and Macaronesia.

ILLUSTRATIONS: Figure 28. Conard, *How to Know the Mosses* (ed. 2), fig. 102a–c. Grout, *Mosses with Hand-Lens and Microscope*, fig. 31. Jennings, *Mosses of Western Pennsylvania* (ed. 2), Pl. 60. Welch, *Mosses of Indiana*, fig. 47.

Lower Peninsula: Clinton Co., *Parmelee*. Gratiot Co., *Schnooberger*. Ingham Co., *Darlington*. Kalamazoo Co., *Becker*. Van Buren Co., *Schnooberger*. Washtenaw Co., *Kauffman*.

2. *Bruchia* Schwaegr.

Very small, gregarious mosses growing from a mostly persistent protonema on soil in open places; leaves subulate from a broader base, mostly longer than the short seta; cells smooth or papillose, rectangular at base; capsules immersed or shortly exserted, obovoid to pear-shaped, erect and symmetric or somewhat inclined, breaking open irregularly; spores maturing in spring or early summer; calyptra mitrate, lobed.

Bruchia sullivantii Aust. — Plants tiny, brownish-green, usually closely gregarious; leaves nearly smooth at apex; costa strong, nearly filling the subula; capsules short-necked, somewhat exserted, about the same length as the setae; spores 30–40 μ, mature in June and July. — On moist soil. Maine to Minnesota and south to the Gulf of Mexico.

ILLUSTRATIONS: Figure 29. Conard, *How to Know the Mosses* (ed. 2), fig. 103a–c. Jennings, *Mosses of Western Pennsylvania* (ed. 2), Pl. 60.

Lower Peninsula: Calhoun Co., *Becker.* Clinton Co., *Parmelee.* Ingham Co., *Darlington.* Kalamazoo Co., *Becker.* Van Buren Co., *Schnooberger.*

3. *Trematodon* Mx.

Yellow-green mosses of moist soil, gregarious or tufted; leaves lanceolate or subulate from a clasping base, ± crisped when dry; costa strong, percurrent; cells smooth and thin-walled; capsules inclined and somewhat curved with a long, slender neck; peristome of 16 teeth; operculum long-rostrate.

Trematodon ambiguus (Hedw.) Hornsch. — Setae bright-yellow, 1–3 cm. long (abundant and making the tufts conspicuous); neck of capsules slender, equal to the urn or somewhat longer; spores in summer. — On moist, clayey or sandy soil in open places. Across northern United States and Canada to Alaska; Europe and Japan.

ILLUSTRATIONS: Figure 30. Grout, *Mosses with Hand-Lens and Microscope,* fig. 38. Jennings, *Mosses of Western Pennsylvania* (ed. 2), Pl. 72.

Upper Peninsula: Chippewa Co., *Steere.* Luce Co., *Gleason, Jr.* Menominee Co., *Hill.*
Lower Peninsula: Eaton Co., *Darlington.*

4. *Distichium* BSG

Densely tufted plants of cooler regions, usually growing in rock crevices; stems forked, radiculose below; leaves 2-ranked, subulate from a sheathing base, rough above; monoicous; setae elongate; capsules cylindric to ovoid, erect or inclined; operculum conic.

Capsules erect . . . . 1. *D. capillaceum*
Capsules inclined . . . . 2. *D. inclinatum*

[40]

1. *Distichium capillaceum* (Hedw.) BSG — Tufts dense, up to 4 or 8 cm. high; upper leaf cells 2–3:1; capsules cylindric, erect; operculum 1–2 mm. long; peristome teeth 16, red; spores ripe in early summer, papillose, 17–20 μ. — Often on rock or calcareous soil. Across Canada and Alaska and southward in the mountains of the United States to South America; Europe, Asia, New Zealand.

ILLUSTRATIONS: Figure 31. Conard. *How to Know the Mosses* (ed. 2), fig. 32a–c. Grout, *Mosses with Hand-Lens and Microscope*, fig. 34 (as *Swartzia montana*).

Upper Peninsula: Alger Co., *Steere.* Keweenaw Co., *Allen & Stuntz.* Mackinac Co., *Wheeler.* Marquette Co., *Hill.*

Lower Peninsula: Alpena Co., *Robinson & Wells.* Cheboygan Co., *Nichols.*

2. *Distichium inclinatum* (Hedw.) BSG — Generally smaller and less silky than the preceding; capsules horizontal; peristome teeth triangular-lanceolate, usually split into 3 divisions; spores warty, 30–45 μ. — On damp calcareous soil or in rock crevices. Across Canada and the northern United States; Europe, northern Africa, and Asia.

ILLUSTRATIONS: Conard, *How to Know the Mosses* (ed. 2), fig. 32d. Grout, *Moss Flora of North America*, vol. 1, Pl. 18c.

Lower Peninsula: Cheboygan Co., *Nichols.* Emmet Co., *Steere.*

5. *Ceratodon* Brid.

Plants in dense tufts; leaves erect-spreading, lanceolate, irregularly toothed above, with recurved margins below; costa strong; cells short-rectangular, thick-walled, smooth; perichaetial leaves differentiated; setae long; capsules inclined to horizontal, asymmetric, sulcate; peristome with 16 bifid teeth; calyptra cucullate.

Ceratodon purpureus (Hedw.) Brid. — The inclined, asymmetric capsules, red-brown and furrowed when old, and the purplish setae are outstanding characters; spores in spring and summer. — One of the most common of all mosses, usually growing in the open on rather dry, often barren substrata. Cosmopolitan and weedy.

ILLUSTRATIONS: Figure 32. Conard, *How to Know the Mosses* (ed. 2), fig. 127. Grout, *Mosses with Hand-Lens and Microscope*, fig. 36. Jennings, *Mosses of Western Pennsylvania* (ed. 2), Pl. 10. Welch, *Mosses of Indiana*, fig. 49.

Upper Peninsula: Chippewa Co., *Thorpe.* Keweenaw Co., *Allen & Stuntz.* Marquette Co., *Nichols.* Ontonagon Co., *Darlington.*

Lower Peninsula: Cheboygan Co., *Nichols.* Clinton Co., *Parmelee.* Eaton Co., *Wheeler.* Ingham and Leelanau counties, *Darlington.* Macomb Co., *Cooley.* Montmorency Co., *Cantlon.*

6. *Saelania* Lindb.

Small, bluish or glaucous, rock- or soil-inhabiting plants of northern regions; stems 1–2 cm. high; leaves lanceolate, slenderly pointed, serrate above; capsules erect and symmetric; peristome teeth 16, bifid to the base, papillose.

Saelania glaucescens (Hedw.) Broth. — Plants small, glaucous or bluish, 1–2 cm. high, densely tufted; upper leaves lanceolate, subulate, strongly costate; cells subquadrate, smooth; capsules cylindric; spores about 15 μ, mature in early autumn. — On the surface or in crevices of calcareous rocks or on moist banks. Greenland; across Canada and the northern United States; Europe, Africa, Asia, Hawaii, and New Zealand.

ILLUSTRATION: Figure 33.

Upper Peninsula: Keweenaw Co., *Cooper.* Ontonagon Co., *Steere.*
Lower Peninsula: Cheboygan Co., *Nichols.* Huron Co., *Davis.*

7. *Trichodon* Schimp.

Plants small, loosely tufted; leaves sheathing at base, abruptly long-subulate, wide-spreading; costa rough at back above, filling the subula; dioicous; setae elongate; capsules suberect or ± inclined, cylindric, smooth; annulus large; peristome teeth bifid nearly to the base, papillose; calyptra cucullate.

Trichodon cylindricus (Hedw.) Schimp. — Upper leaves squarrose, contorted when dry, subulate, rough above; capsules erect and symmetric, narrowly cylindric; spores 11–14 μ, smooth. — On soil. Alaska to Idaho; Michigan; Europe and Asia.

ILLUSTRATIONS: Figure 34. Grout. *Moss Flora of North America*, vol. 1, Pl. 33 (as *Ditrichum*).

Upper Peninsula: Keweenaw Co. (Mt. Bohemia), *Ireland.*

8. *Ditrichum* Hampe

Mostly small, loosely tufted plants growing on soil or rock, usually in the open; leaves lance-subulate or awned, broadly costate; upper cells subquadrate to elongate, smooth; capsules ± erect, symmetric or somewhat asymmetric, often slightly furrowed; annulus large; peristome teeth divided nearly to the base, papillose.

1. Plants tall (5–10 cm. high), brown-tomentose . . 3. *D. flexicaule*
1. Plants short (usually less than 1.5 cm. high, not
    tomentose) 2

2. Setae bright-yellow . . . . 4. *D. pallidum*
2. Setae red-brown 3
3. Capsules 0.5 – 1.5 mm. long, usually smooth when dry;
   awn of perichætial leaves as long as the base or longer;
   innovations with imbricate leaves none . . . . 1. *D. pusillum*
3. Capsules about 1.5 mm. long, often wrinkled when dry;
   awn of perichætial leaves much shorter than the base;
   innovations with imbricate leaves common . . . . 2. *D. lineare*

1. *Ditrichum pusillum* (Hedw.) E. Britt. — Small plants in thin sods, without differentiated innovations; setae and capsules red-brown; spores in autumn and winter. — On bare soil. Widespread in eastern North America; British Columbia; Europe, northern Africa, and Asia.

ILLUSTRATIONS: Conard, *How to Know the Mosses* (ed. 2), fig. 108a–d. Jennings, *Mosses of Western Pennsylvania* (ed. 2), Pl. 10. Welch, *Mosses of Indiana*, fig. 50.

Lower Peninsula: Clinton Co., *Parmelee*. Kalamazoo Co., *Becker*.

2. *Ditrichum lineare* (Sw.) Lindb. — Similar to the preceding but bearing terete innovations with blunt, appressed leaves. — On bare soil. Widespread in eastern North America; Europe.

ILLUSTRATIONS: Conard, *How to Know the Mosses* (ed. 2), fig. 108e. Jennings, *Mosses of Western Pennsylvania* (ed. 2), Pl. 9.

Upper Peninsula: Alger Co., *Nichols*.

3. *Ditrichum flexicaule* (Schwaegr.) Hampe — Plants in dense, dull, green or yellow-green tufts; leaves erect-spreading, flexuous or contorted when dry, sometimes twisted at apex, gradually subulate from a lanceolate base; cells at leaf shoulders paler and obliquely short-rhomboidal forming a suggestion of a border; setae dark red, 10–25 mm. long; spores in spring. — On moist soil or rocks, particularly on banks and cliffs. Greenland; across the continent, south to Idaho, Colorado, and Vermont; Europe and Asia.

ILLUSTRATIONS: Figure 35. Grout, *Moss Flora of North America*, vol. 1, Pl. 29A, 30A.

Upper Peninsula: Mackinac Co., *Nichols*.
Lower Peninsula: Emmet Co., *Steere*.

4. *Ditrichum pallidum* (Hedw.) Hampe — Plants in loose, yellowish-green tufts; leaves subulate, channeled; costa strong, long-excurrent, basal cells oblong-hexagonal; setae long, bright-yellow; capsules ± inclined and somewhat asymmetric; operculum bluntly

[43]

conic; peristome teeth reddish, ± completely divided; spores in early summer. — Common on dry, sandy soil. Widespread in eastern North America; Europe, Africa, Azores, and Japan.

ILLUSTRATIONS: Conard, *How to Know the Mosses* (ed. 2), fig. 129. Grout, *Mosses with Hand-Lens and Microscope*, fig. 33. Jennings, *Mosses of Western Pennsylvania* (ed. 2), Pl. 10. Welch, *Mosses of Indiana*, fig. 51.

Upper Peninsula: Chippewa Co., *Steere.*
Lower Peninsula: Jackson Co., *Marshall.* Oakland Co., *Darlington.* Van Buren Co., *Kauffman.*

SELIGERIACEAE

Mostly very small, rock-inhabiting plants; stems erect, simple or branched; leaves lanceolate or subulate, costate; cells smooth; setae elongate, sometimes curved; capsules erect, ovoid or shortly pyriform, often turbinate when dry; operculum differentiated; peristome usually present, consisting of 16 mostly entire teeth.

Plants very small (up to 5 mm. high); alar cells
  not enlarged . . . . 1. *Seligeria*
Plants larger; alar cells conspicuously enlarged . . . . . . 2. *Blindia*

1. *Seligeria* BSG

Plants minute, gregarious or loosely tufted; leaves sheathing at base; alar cells not or only slightly differentiated; setae elongate, often curved; capsules erect, symmetric, ovoid to shortly pyriform; operculum rostrate; annulus none; peristome of 16 teeth (or rarely lacking); calyptra cucullate.

1. Peristome lacking . . . . 1. *S. doniana*
1. Peristome present 2
  2. Setae straight or nearly so; capsules broadest at
      the mouth, with a short, distinct neck 3
  2. Setae curved when moist; capsules not broadest
      at the mouth 4
3. Leaves tapered to a long, slender awn . . . . . 3. *S. pusilla*
3. Leaves abruptly narrowed to a short, thick awn . . 2. *S. calcarea*
  4. Leaves entire; costa filling the long awn . . . 4. *S. recurvata*
  4. Leaves often serrulate; costa not filling
      the short subula . . . . 5. *S. campylopoda*

1. *Seligeria doniana* (Sm.) C.M. — Tiny, dark-green plants (1 mm. or less high); stems simple or forked; leaves subulate from a broader, serrulate base; setae 2 mm. long; capsules pyriform; peristome none; operculum conic-rostrate; spores 8–10 μ, maturing in late summer. — A rare moss, on calcareous rocks. Canada and the eastern United States; Europe and Asia.

[44]

ILLUSTRATION: Grout, *Moss Flora of North America*, vol. 1, Pl. 35.

Upper Peninsula: Mackinac Co., *Steere*.
Lower Peninsula: Alpena Co., *Robinson* & *Wells*.

2. *Seligeria calcarea* (Dicks.) BSG — Leaves abruptly narrowed to a short, thick acumen, subentire; peristome red-brown. — On wet limestone. Ontario to Minnesota, south to Missouri; Europe.

ILLUSTRATIONS: Figure 36. Conard, *How to Know the Mosses* (ed. 2), fig. 107e.

Lower Peninsula: Alpena Co., *Robinson* & *Wells*. Emmet Co., *Steere*.

3. *Seligeria pusilla* (Hedw.) BSG — Tufts small (up to 1 mm. high), dark-green; upper leaves filiform-subulate, about 2 mm. long, with incurved, subdenticulate margins; setae 1.5–3 mm. long; capsules obovoid; peristome teeth broadly lanceolate, entire, smooth; spores 10–14 μ, maturing in summer. — On damp, shaded cliffs and ledges, especially limestone. Ontario and Minnesota, south to Missouri; Europe and Asia.

ILLUSTRATIONS: Conard, *How to Know the Mosses* (ed. 2), fig. 107d. Grout, *Moss Flora of North America*, vol. 1, Pl. 34.

Upper Peninsula: Alger Co. (Whitefish Point), *Beck*.

4. *Seligeria recurvata* (Hedw.) BSG — Tufts small (1–3 mm. high), dark-green; upper leaves subulate from a lanceolate base, entire; setae 3–4 mm. long, curved when moist; capsules ovoid, up to 0.75 mm. long; peristome teeth lanceolate (with about 10 joints); operculum long-rostrate; spores about 10 μ, maturing in late spring or early summer. — Rare, on damp rocks (probably preferring sandstone). Eastern North America; British Columbia and Washington; Europe.

ILLUSTRATION: Grout, *Moss Flora of North America*, vol. 1, Pl. 35C.

Upper Peninsula: Marquette Co. (Pine River Point), *Nichols*.

5. *Seligeria campylopoda* Lindb. — Plants gregarious, up to 1 mm. high; leaves oblong-lanceolate, often serrulate, 0.75–1.5 mm. long; costa stout, mostly ending below the apex; setae curved when moist; capsules ovoid; peristome teeth lanceolate (with 12–14 joints); spores in autumn. — On damp, shaded limestone; uncommon. Northeastern United States and adjacent provinces of Canada; British Columbia, Alberta, and Montana; northern Europe.

[45]

ILLUSTRATIONS: Conard, *How to Know the Mosses* (ed. 2), fig. 107a–c. Grout, *Moss Flora of North America*, vol. 1, Pl. 35A.

Lower Peninsula: Cheboygan Co., *Nichols.* Emmet Co., *Steere.* Mackinac Co., *Nichols.*

2. *Blindia* BSG

Plants gregarious or tufted, growing on wet, often wave-washed bluffs and ledges; leaves lanceolate, auriculate at base; alar cells inflated; capsules globose-pyriform, often turbinate when dry; peristome of 16 entire or perforate teeth.

Blindia acuta (Hedw.) BSG — Plants up to 8 cm. high, usually less; leaves suberect, lance-subulate from a sheathing base, 2–3 mm. long, usually deciduous below; margins entire, incurved; alar cells quadrate or short-rectangular, ± colored; setae 3–10 mm. long; capsules turbinate when dry; spores 10–18 μ. — On wet rocks, particularly along brooks. Greenland to Alaska, south to North Carolina, Colorado, and California; Europe, Asia, Azores, and Madeira.

ILLUSTRATIONS: Figure 37. Grout, *Mosses with Hand-Lens and Microscope*, fig. 41.

Upper Peninsula: Alger Co., *Steere.* Delta Co., *Wheeler.* Keweenaw Co., *Steere.* Marquette Co., *Nichols.*

DICRANACEAE

Plants erect, tufted, often tomentose. Leaves narrow, often sheathing at base, frequently secund; costa well developed, seldom ending below the apex; basal cells elongate, upper short or long, alar often well marked; setae usually long; capsules erect or inclined; operculum mostly rostrate; peristome teeth 16, mostly deeply bifid; calyptra cucullate.

1. Leaves ± falcate-secund 2
1. Leaves usually not falcate-secund 3
   2. Alar cells conspicuously differentiated; mostly ± robust plants . . . . 7. *Dicranum*
   2. Alar cells not differentiated; small plants . . . 1. *Dicranella*
3. Leaves setaceous from a lanceolate base; alar cells enlarged and often colored . . . . . 8. *Dicranodontium*
3. Leaves not setaceous; alar cells scarcely differentiated 4
   4. Leaves ovate-lanceolate or lingulate; cells rounded-quadrate and mammillose above and in several rows at the basal margins . . . . 4. *Dichodontium*
   4. Leaves not as above 5
5. Capsules erect to horizontal, ± curved or asymmetric, sometimes strumose . . . . 5. *Oncophorus*
5. Capsules erect and symmetric, not strumose 6

6. Leaves subulate from a lanceolate base, channeled
above, entire; setae 6–10 mm. long . . . 6. *Dicranoweisia*
6. Leaves linear-lanceolate or narrowly lingulate,
serrulate; setae 2–5 mm long 7
7. Capsules plicate when dry; leaf cells minutely
papillose . . . . 2. *Rhabdoweisia*
7. Capsules smooth; leaf cells mammillose . . . . 3. *Oreoweisia*

1. *Dicranella* (C.M.) Schimp.

Plants small, gregarious or tufted; leaves narrow, often secund;
costa well developed; alar cells not differentiated; capsules erect
or inclined, sometimes asymmetric; operculum ± beaked.

1. Leaves erect-spreading or somewhat secund 2
1. Leaves wide-spreading or squarrose from an erect,
± sheathing base 4
2. Setae yellow; plants usually more than
1 cm. high . . . . 5. *D. heteromalla*
2. Setae reddish; stems usually less than 1 cm. high 3
3. Capsules smooth; operculum short-rostrate . . . . 3. *D. varia*
3. Capsules plicate when dry and empty; operculum
subulate-rostrate . . . . 4. *D. subulata*
4. Leaves toothed above; costa toothed at back;
seta yellowish . . . . 1. *D. schreberiana*
4. Leaves entire or nearly so; costa smooth;
seta red . . . . 2. *D. grevilleana*

1. *Dicranella schreberiana* (Hedw.) Schimp. — Tufts yellow-
green, 1–3 cm. high; leaves lanceolate and squarrose from a sheathing base, toothed above; costa ± toothed at back; capsule 0.8–1.4
mm. long; operculum rostrate; spores in autumn. — On damp clay
in open places. New England to British Columbia; Europe, Asia,
and New Zealand.

ILLUSTRATIONS: Conard, *How to Know the Mosses* (ed. 2), fig. 132. Grout,
Moss Flora of North America, vol. 1, Pl. 37E.

Upper Peninsula: Keweenaw Co., *Holt.* Luce Co., *Nichols.* Ontonagon Co.,
Nichols & Steere.
Lower Peninsula: Cheboygan Co., *Nichols.* Montmorency Co., *Cantlon.*
Presque Isle Co., *Steere.*

2. *Dicranella grevilleana* (Brid.) Schimp. — Leaves lance-
subulate and squarrose from a sheathing base, entire or nearly so;
costa smooth at back; operculum rostrate; spores maturing in
summer. — On damp clay or rock ledges. Across the continent,
south to Michigan and Washington; Europe and Asia.

ILLUSTRATION: Grout, *Moss Flora of North America*, vol. 1, Pl. 37D.

Upper Peninsula: Alger Co., *Wheeler.* Keweenaw Co., *Steere.* Mackinac
and Marquette counties, *Nichols.*

[47]

3. *Dicranella varia* (Hedw.) Schimp. — Plants dull- to bright-green. Leaves 1.5–2 mm. long, narrowly lanceolate; margins recurved below; cells short-rectangular above; capsules curved and inclined, smooth; spores maturing in late autumn and winter. — On moist clay in open places. Across Canada and Alaska, south to California in the West, Florida in the East; Europe, North Africa, Macaronesia, and Asia.

ILLUSTRATIONS: Grout, *Mosses with Hand-Lens and Microscope*, Pl. 13. Jennings, *Mosses of Western Pennsylvania* (ed. 2), Pl. 11. Welch, *Mosses of Indiana*, fig. 52.

Upper Peninsula: Mackinac Co., *Nichols*.
Lower Peninsula: Barry Co., *Becker*.

4. *Dicranella subulata* (Hedw.) Schimp. — Plants in low tufts; leaves 2–3 mm. long, subulate from a lanceolate base; upper cells linear, 3–4 μ wide; capsules ± curved, plicate when dry and empty; spores in late summer or autumn. — Usually on rock ledges or stony soil. Greenland; northern United States, far north in Canada; Europe and Asia.

ILLUSTRATION: Grout, *Moss Flora of North America*, vol. 1, Pl. 37E (as *D. secunda*).

Upper Peninsula: Houghton Co., *Steere & Sharp*. Marquette Co., *Nichols*.

5. *Dicranella heteromalla* (Hedw.) Schimp. — Leaves lance-subulate, often falcate-secund, serrulate above; cells short-rectangular; capsules 1–1.5 mm. long, slightly curved, sulcate when dry and empty, the mouth large and oblique; operculum long-rostrate; spores in autumn or winter. — On moist, shaded banks. Across the continent, south to the Gulf of Mexico; Europe, Macaronesia, and Asia.

ILLUSTRATIONS: Figure 38. Conard, *How to Know the Mosses* (ed. 2), fig. 131. Grout, *Mosses with Hand-lens and Microscope*, Pl. 11. Jennings, *Mosses of Western Pennsylvania* (ed. 2), Pl. 11. Welch, *Mosses of Indiana*, fig. 53.

Upper Peninsula: Chippewa Co., *Steere*. Houghton Co., *Parmelee*. Keweenaw Co., *Gilly & Parmelee*. Luce Co., *Nichols*. Ontonagon Co., *Nichols & Steere*.
Lower Peninsula: Alpena Co., *Robinson & Wells*. Cheboygan Co., *Nichols*. Eaton, Ingham, and Leelanau counties, *Darlington*. Washtenaw Co., *Kauffman*.

2. Rhabdoweisia BSG

Small, tufted plants growing on rock; leaves linear-lanceolate, acute, spreading, crisped when dry; costa ending slightly below the apex; cells green and rounded-quadrate above, pale and rectangular below; autoicous; setae yellow; capsules symmetric, short-oblong, 8-plicate; peristome teeth 16, undivided.

Rhabdoweisia denticulata (Brid.) BSG — Small, densely tufted plants (less than 1 cm. high); leaves oblong- to linear-lanceolate, serrulate above; upper cells finely papillose; capsules small, erect or nearly so, plicate; spores in summer. — In moist rock crevices. Newfoundland to Alaska and south to North Carolina; Europe and Hawaii.

ILLUSTRATIONS: Figure 39. Grout, *Moss Flora of North America*, vol. 1, Pl. 39B. Jennings, *Mosses of Western Pennsylvania* (ed. 2), Pl. 11.

Upper Peninsula: Alger Co., *Steere.* Luce Co., *Nichols.* Menominee Co., *Hill.*

3. *Oreoweisia* (Schimp.) DeNot.

Plants small, mostly restricted to shaded ledges in more elevated regions. Leaves narrowly lingulate; costa ending below the apex, with 1 stereid band; cells rounded-quadrate and mammillose above; pale, smooth, and rectangular below; capsules exserted, oval, erect or slightly inclined, smooth.

Oreoweisia serrulata (Funck) DeNot. — Tufts small and compact, 1–6 mm. high; leaves spreading from an erect base, incurved when dry, keeled, serrulate; costa strong. — On rock. Across the continent, south to Tennessee; Europe.

ILLUSTRATIONS: Figure 40. Grout, *Moss Flora of North America*, vol. 1, Pl. 40A.

Upper Peninsula: Keweenaw Co., *Hermann.*

4. *Dichodontium* Schimp.

Plants tufted, yellow-green, radiculose below, on rocks and soil in cool regions; leaves lanceolate or lingulate and spreading from a broader, erect base, incurved-crisped when dry; costa strong, with 2 stereid bands; cells rounded-quadrate and mammillose above and at basal margins; capsules exserted, oblong-ovoid, ± inclined; peristome of 16 teeth.

Dichodontium pellucidum (Hedw.) Schimp. — Leaves soft, irregularly serrate above; brood-bodies sometimes borne in leaf axils. — Across the continent and south to Pennsylvania, the Great Lakes, and California; Europe, Madeira, and Asia.

[49]

ILLUSTRATION: Figure 41.

Upper Peninsula: Alger Co., *Nichols.* Keweenaw Co., *Steere.* Luce and Marquette counties, *Nichols.* Ontonagon Co., *Nichols & Steere.*

Dichodontium pellucidum var. *fagimontanum* (Brid.) Schimp. — A slender form with relatively short, broad leaves. — Occasional in North America and Europe.

Upper Peninsula: Alger Co. (Grand I.), *Nichols.*

5. *Oncophorus* Brid.

Tufted plants of elevated or northern regions, usually on soil or rocks, either in the open or in somewhat shaded situations; stems forked above, closely foliate; leaves often widely spreading, mostly crisped when dry, lanceolate; costa well developed; cells small, opaque; capsules exserted, usually asymmetric, sometimes strumose; operculum beaked; peristome teeth 16, mostly papillose and divided.

1. Capsules strumose 2
1. Capsules not strumose 4
  2. Leaves strongly papillose above; capsules ribbed
    when dry . . . . 2. *O. strumiferus*
  2. Leaves faintly or not at all papillose; capsules
    not ribbed 3
3. Leaves keeled above; margins recurved in part . . . 5. *O. virens*
3. Leaves not keeled; margins mostly plane . . . 6. *O. wahlenbergii*
  4. Leaves lanceolate or ovate-lanceolate; tufts
    up to 5 cm. high . . . . 1. *O. polycarpus*
  4. Leaves linear-lanceolate; tufts mostly less than 2 cm. high . . . 5
5. Peristome teeth split 2/3 down . . . . . . 3. *O. tenellus*
5. Peristome teeth not divided . . . . . . 4. *O. schisti*

1. *Oncophorus polycarpus* (Hedw.) Brid. — Leaves lanceolate to ovate-lanceolate; margins mostly bistratose, irregularly serrate above; setae straight, yellowish; capsules erect or nearly so, furrowed when dry; operculum obliquely beaked; spores in summer. — On moist surfaces, especially rock ledges. Great Lakes to Montana and northward to Greenland and Alaska; Europe and Asia.

ILLUSTRATION: Grout, *Moss Flora of North America*, vol. 1, Pl. 42E.

Upper Peninsula: Keweenaw Co., *Cooper.* Marquette Co., *Nichols.* Ontonagon Co., *Nichols & Steere.*

2. *Oncophorus strumiferus* Brid. — Differing from the preceding mainly in the strumose, ± curved and inclined capsules and the unistratose leaf margins. — On shaded rocks. New England to Montana and northward; Europe and Asia.

ILLUSTRATIONS: Dixon, *Handbook of British Mosses*, Pl. 9N (as *O. polycarpus* var.). Grout, *Moss Flora of North America*, vol. 1, Pl. 42C.

Upper Peninsula: Keweenaw Co., *Cooper.* Marquette Co., *Nichols.*

3. *Oncophorus tenellus* (BSG) Williams — Leaves linear-lanceolate; peristome teeth mostly split below the middle; beak of operculum 1.5–2 times the length of the conic base. — On moist rock ledges. New England to Montana and north to Alaska; Europe and Asia.

ILLUSTRATION: Grout, *Moss Flora of North America*, vol. 1, Pl. 41B.

Upper Peninsula: Keweenaw Co., *Povah.* Marquette Co., *Nichols.*

4. *Oncophorus schisti* (Wahl.) Lindb. — Upper leaves linear-lanceolate, keeled; costa stout, ending below the apex; capsules erect, up to 1 mm. long; peristome teeth not split; spores in spring. — Usually on moist rock surfaces or on soil in the crevices of cliffs. Michigan to Montana and north to Alaska and Greenland; Europe and Asia.

ILLUSTRATION: Grout, *Moss Flora of North America*, vol. 1, Pl. 41B.

Upper Peninsula: Keweenaw Co., *Steere.*

5. *Oncophorus virens* (Hedw.) Brid. — Plants closely tufted, radiculose below; upper leaves linear-lanceolate from a larger, subclasping base, up to 1.8 mm. long, keeled; margins ± recurved; costa stout, ending below the apex; cells smooth; capsules curved and asymmetric, strumose; peristome teeth reddish, narrowly lanceolate; spores in spring. — On moist soil or rock. Across the continent, south to Michigan and California; Europe, northern Africa, and Asia.

ILLUSTRATION: Grout, *Moss Flora of North America*, vol. 1, Pl. 42.

Upper Peninsula: Alger and Keweenaw counties, *Steere.*

6. *Oncophorus wahlenbergii* Brid. — Plants bright-green or yellowish, densely tufted; leaves linear-lanceolate from a strongly sheathing base, subtubulose above, wide-spreading and strongly crisped when dry; margins not recurved; costa often excurrent; capsules strumose; operculum beaked; spores in spring. — On decaying logs or soil in shade. Across Canada to Alaska and south to the northern United States; Europe and Asia.

[51]

ILLUSTRATIONS: Figure 42. Conard, *How to Know the Mosses* (ed. 2), fig. 115. Grout, *Mosses with Hand-Lens and Microscope*, fig. 35. Jennings, *Mosses of Western Pennsylvania* (ed. 2), Pl. 12.

Upper Peninsula: Chippewa and Keweenaw counties, *Steere*. Luce and Marquette counties, *Nichols*. Ontonagon Co., *Darlington & Bessey*.

Lower Peninsula: Alpena Co., *Robinson & Wells*. Cheboygan Co., *Nichols*. Clare Co., *Schnooberger*. Emmet Co., *Steere*. Leelanau Co., *Darlington*. Montcalm Co., *Schnooberger*.

6. *Dicranoweisia* Lindb.

Low, densely tufted plants; leaves subulate from a lanceolate base, crisped when dry, entire; costa ending in the apex; upper cells subquadrate or shortly rectangular, smooth or slightly papillose; lower cells elongate, not or slightly differentiated at the basal angles; autoicous; setae elongate; capsules erect and symmetric, smooth; peristome teeth entire or split at the apex; operculum long-rostrate.

Dicranoweisia crispula (Hedw.) Lindb. — Plants up to 1 or 2 cm. high, yellow-green; leaves up to 4 mm. long, lance-linear, spreading and crisped when dry, with margins erect or incurved; inner perichaetial leaves rounded-obtuse to abruptly short-acuminate; capsules cylindric; annulus none; spores mature in early spring. — On non-calcareous rocks in cooler regions. Greenland; Canada and the northern United States; Europe and Asia.

ILLUSTRATIONS: Figure 43. Conard, *How to Know the Mosses* (ed. 2), fig. 112e.

Upper Peninsula: Keweenaw Co. (Copper Harbor; Manganese R.), *Steere*.

7. *Dicranum* Hedw.

Plants small to large, densely tufted, usually tomentose; leaves mostly lance-acuminate, often falcate-secund; costa well developed; cells short or elongate, often porose, mostly smooth, alar cells differentiated in well-marked, often colored groups; setae elongate; capsules cylindric, suberect or more often curved and asymmetric; peristome teeth deeply forked; operculum long-rostrate; calyptra cucullate.

1. Plants robust (usually more than 7 cm. high); capsules
   ± curved and inclined 2
1. Plants small to medium-sized (usually less than 6 cm. high);
   capsules various 7
   2. Upper cells short (1–2:1, rarely more), not pitted 3
   2. Upper cells longer, ± pitted 5
3. Leaves not rugose or secund . . . . . . 6. *D. muehlenbeckii*
3. Leaves rugose or undulate 4

[52]

1. *Dicranum montanum* Hedw. — Small, yellow-green plants; leaves subtubulose, strongly crisped; costa about 1/5 the width of the leaf base; upper cells short, ± papillose; capsules suberect. — On bark, especially at the base of trees. Widespread in eastern North America; Arizona; Europe and Asia.

ILLUSTRATIONS: Grout, *Moss Flora of North America*, vol. 1, Pl. 44. Jennings, *Mosses of Western Pennsylvania* (ed. 2), Pl. 12. Welch, *Mosses of Indiana*, fig. 55.

Upper Peninsula: Alger Co., *Gilly & Parmelee*. Chippewa Co., *Steere*. Houghton Co., *Parmelee*. Keweenaw Co., *Hermann*. Mackinac Co., *Gilly & Parmelee*. Marquette Co., *Nichols*. Ontonagon Co., *Darlington*.
Lower Peninsula: Alpena Co., *Hall*. Cheboygan Co., *Nichols*. Eaton Co., *Darlington*. Emmet Co., *Gilly & Parmelee*. Gratiot and Huron counties, *Schnooberger*. Leelanau Co., *Darlington*.

2. *Dicranum flagellare* Hedw. — Sterile plants commonly bearing flagelliform branches in upper leaf axils; leaves lanceolate, 3-4

[53]

mm. long, subtubulose, ± secund and crisped when dry; costa sub-percurrent; upper cells short, smooth; capsules suberect; spores in late summer. — Usually on rotten wood or peaty soil in shady places. Nova Scotia to British Columbia, south to South Carolina; Mexico; Europe and Asia.

ILLUSTRATIONS: Conard, *How to Know the Mosses* (ed. 2), fig. 119. Grout, *Mosses with Hand-Lens and Microscope*, fig. 46c, c', d. Jennings, *Mosses of Western Pennsylvania* (ed. 2), Pl. 12. Welch, *Mosses of Indiana*, fig. 56.

Upper Peninsula: Chippewa Co., *Steere*. Houghton Co., *Parmelee*. Keweenaw Co., *Allen & Stuntz*. Marquette Co., *Nichols*. Ontonagon Co., *Nichols & Steere*.

Lower Peninsula: Alpena Co., *Hall*. Cheboygan Co., *Nichols*. Genesee Co., *Darlington*. Ionia Co., *Parmelee*. Kalamazoo, Leelanau, and Oakland counties, *Darlington*. Oscoda Co., *Parmelee*. St. Joseph Co., *Darlington*. Washtenaw Co., *Kauffman*.

3. *Dicranum fulvum* Hook. — Leaves erect-spreading or falcate-secund, long-lanceolate and gradually subulate, indistinctly serrulate near the apex; costa excurrent, about 1/3 the width of the leaf base; upper cells short, rather obscure; capsules erect; spores in autumn. — On shaded rocks. Widespread in eastern North America; Europe.

ILLUSTRATIONS: Grout, *Mosses with Hand-Lens and Microscope*, fig. 46e, e'. Jennings, *Mosses of Western Pennsylvania* (ed. 2), Pl. 13. Welch, *Mosses of Indiana*, fig. 57.

Upper Peninsula: Houghton Co., *Hyypio*.
Lower Peninsula: Washtenaw Co., *Steere*.

4. *Dicranum viride* (Sull. & Lesq.) Lindb. — Smaller than *D. fulvum*, with leaves entire, fragile, not or only slightly secund. — On logs and tree trunks in woods. Southern Canada and northern United States, south to Pennsylvania and Ohio; Europe and Asia.

ILLUSTRATIONS: Grout, *Mosses with Hand-Lens and Microscope*, fig. 48. Jennings, *Mosses of Western Pennsylvania* (ed. 2), Pl. 13 (as *D. fulvum* var.). Welch, *Mosses of Indiana*, fig. 58.

Upper Peninsula: Chippewa Co., *Steere*. Gogebic Co., *Golley*. Marquette Co., *Nichols*. Ontonagon Co., *Nichols & Steere*.
Lower Peninsula: Alpena Co., *Robinson & Wells*. Cheboygan Co., *Nichols*. Leelanau Co., *Darlington*.

5. *Dicranum longifolium* Hedw. — Plants usually grayish- or bluish-green; leaves long-subulate, usually falcate-secund, denticulate; costa about 1/2 or more the width of the leaf base, consisting of 3–4 layers of empty, hyaline cells enclosing a layer of small, green cells. — On boulders, rocky ledges, or tree trunks. Greenland; New-

foundland to Alaska, south to North Carolina and Colorado; Europe and Asia.

ILLUSTRATIONS: Grout, *Mosses with Hand-Lens and Microscope*, Pl. 17. Jennings, *Mosses of Western Pennsylvania* (ed. 2), Pl. 13.

Upper Peninsula: Chippewa Co., *Thorpe*. Houghton Co., *Hyypio*. Keweenaw and Mackinac counties, *Steere*. Marquette Co., *Nichols*. Ontonagon Co., *Darlington & Bessey*.
Lower Peninsula: Emmet Co., *Steere*.

6. *Dicranum muehlenbeckii* BSG — Similar to *D. fuscescens* but differing in having subtubulose leaves with unistratose margins; spores in summer to autumn. — On soil, rocks, and decaying logs. Across the continent, south to New Jersey, the Great Lakes, and New Mexico; Europe and Asia.

ILLUSTRATION: Grout, *Mosses with Hand-Lens and Microscope*, Pl. 16.

Upper Peninsula: Houghton Co., *Parmelee*. Ontonagon Co., *Nichols & Steere*.
Lower Peninsula: Charlevoix Co., *Parmelee*. Cheboygan Co., *Nichols*.

7. *Dicranum fuscescens* Turn. — Plants extremely variable but usually easily recognized by its leaves crisped and often secund, strongly keeled, and bistratose here and there along the margins; cells short; capsules curved and inclined. — On soil, rocks, and decayed wood. Across the continent, south to South Carolina and California; Europe and Asia.

ILLUSTRATIONS: Conard, *How to Know the Mosses* (ed. 2), fig. 59a–d, 120a–d. Grout, *Moss Flora of North America*, vol. 1, Pl. 46.

Upper Peninsula: Chippewa Co., *Steere*. Keweenaw Co., *Hermann*. Ontonagon Co., *Nichols & Steere*.
Lower Peninsula: Cheboygan Co., *Nichols*. Huron Co., *Kauffman*.

8. *Dicranum bergeri* Bland. — Robust plants in dense, deep, tomentose clumps; leaves erect, little changed on drying, ± undulate, broadly pointed; upper cells about 1–3:1; setae single; capsules curved and inclined; spores in summer. — In bogs in cooler regions. Greenland; across the continent and south to New Jersey, the Great Lakes, and Colorado; Europe and Asia.

ILLUSTRATIONS: Figure 44. Grout, *Moss Flora of North America*, vol. 1, Pl. 46B.

Upper Peninsula: Chippewa Co., *Steere*. Keweenaw Co., *Hermann*. Luce Co., *Nichols*.
Lower Peninsula: Cheboygan Co., *Nichols*. Washtenaw Co., *Steere*.

9. *Dicranum drummondii* C.M. — Rather robust plants; leaves slenderly pointed, crisped when dry, ± rugose; setae clustered; spores in summer. — On cool, moist substrata, in woods. Canada and the United States south to New Jersey and Colorado.

ILLUSTRATION: Grout, *Moss Flora of North America*, vol. 1, Pl. 48A, 50C.

Upper Peninsula: Alger Co., *Wheeler*. Chippewa Co., *Steere*. Houghton Co., *Parmelee*. Keweenaw Co., *Allen & Stuntz*. Marquette Co., *Nichols*. Ontonagon Co., *Nichols & Steere*.
Lower Peninsula: Alpena and Charlevoix counties, *Wheeler*. Cheboygan Co., *Nichols*.

10. Dicranum condensatum Hedw. — Plants green or yellowish, radiculose, about 2–4 cm. high; leaves erect-spreading when moist, curved or crisped when dry, oblong-lanceolate, serrulate toward the apex; costa percurrent or shortly excurrent; cells short, smooth or ± papillose at back; setae single; capsules curved and inclined. — Usually on sand. Widespread in eastern North America.

ILLUSTRATIONS: Conard, *How to Know the Mosses* (ed. 2), fig. 59e–f, 120e–f. Grout, *Mosses with Hand-Lens and Microscope*, fig. 45c, c'. Welch, *Mosses of Indiana*, fig. 59.

Upper Peninsula: Alger Co., *Parmelee*. Marquette Co., *Nichols*.
Lower Peninsula: Cheboygan Co., *Nichols*. Emmet Co., *Steere*.

11. Dicranum spurium Hedw. — Short but stout, yellow-green plants (up to 2 or 4 cm. high); leaves ovate, undulate, serrulate, erect and somewhat crisped when dry; costa subpercurrent; upper cells irregular, triangular and quadrate, mammillose at back; capsules curved and inclined; spores in spring. — On sand or rocks. Newfoundland to the Great Lakes, south to Tennessee and Missouri; Europe and Asia.

ILLUSTRATION: Grout, *Mosses with Hand-Lens and Microscope*, fig. 45d–e.

Upper Peninsula: Marquette Co., *Nichols*. Ontonagon Co., *Nichols & Steere*.
Lower Peninsula: Cheboygan Co., *Nichols*. Emmet and Washtenaw counties, *Steere*.

12. Dicranum fragilifolium Lindb. — Leaves erect-spreading or slightly secund, slenderly subulate from a lanceolate base, fragile; costa long-excurrent; upper cells quadrate or rectangular; capsules curved and inclined. — On decaying stumps and logs. Labrador to Alaska, south to the Great Lakes and Oregon.

Upper Peninsula: Keweenaw Co. (Isle Royale), *Allen & Stuntz*.

13. Dicranum rugosum Brid. — Robust, glossy, green or yellowish, tomentose plants in loose tufts; leaves large, long-lanceolate, wide-spreading and rugose; costa narrow; upper cells elongate, porose; setae clustered; capsules curved and inclined; spores in late summer. — On soil or humus in woods. Canada and the northern United States, south to Ohio and Virginia; Europe and Asia.

ILLUSTRATIONS: Grout, *Mosses with Hand-Lens and Microscope*, fig. 46a (as *D. undulatum*). Jennings, *Mosses of Western Pennsylvania* (ed. 2), Pl. 61.

Upper Peninsula: Chippewa Co., *Steere*. Houghton Co., *Parmelee*. Keweenaw Co., *Hermann*. Mackinac and Marquette counties, *Parmelee*. Menominee Co., *Hill*. Ontonogan Co., *Darlington & Bessey*.

Lower Peninsula: Alpena Co., *Wheeler*. Charlevoix Co., *Parmelee*. Cheboygan Co., *Nichols*. Crawford, Emmet, and Ingham counties, *Parmelee*. Iosco Co., *Cantlon*. Leelanau Co., *Darlington*. Montmorency Co., *Cantlon*. Van Buren Co., *Kauffman*. Washtenaw Co., *Steere*.

14. *Dicranum scoparium* Hedw. — Robust, yellow-green plants; leaves falcate-secund, subulate, strongly serrate; costa subpercurrent o short-excurrent; upper cells elongate, porose; setae single; capsules curved and inclined. — On soil, logs, and rocks in woods. Widespread across the continent and south to the Gulf of Mexico and California; Europe, Asia, Macaronesia, and New Zealand.

ILLUSTRATIONS: Conard, *How to Know the Mosses* (ed. 2), fig. 117a–d. Grout, *Mosses with Hand-Lens and Microscope*, fig. 42, 43. Jennings, *Mosses of Western Pennsylvania* (ed. 2), Pl. 12. Welch, *Mosses of Indiana*, fig. 60.

Upper Peninsula: Alger Co., *Wheeler*. Baraga Co., *Parmelee*. Chippewa Co., *Thorpe*. Delta Co., *Wheeler*. Keweenaw and Mackinac counties, *Parmelee*. Marquette Co., *Nichols*. Menominee Co., *Hill*. Ontonogan Co., *Darlington & Bessey*.

Lower Peninsula: Alpena Co., *Hall*. Cheboygan Co., *Nichols*. Clinton Co., *Parmelee*. Eaton and Emmet counties, *Wheeler*. Leelanau Co., *Darlington*. Macomb Co., *Cooley*.

15. *Dicranum bonjeanii* DeNot. — Differing from *D. scoparium* in the loosely-erect leaves which are broadly pointed and slightly rugose above. — On damp substrata in woods. Widespread in North America; Europe, Azores, and Asia.

ILLUSTRATIONS: Conard, *How to Know the Mosses* (ed. 2), fig. 117e. Grout, *Mosses with Hand-Lens and Microscope*, Pl. 15. Welch, *Mosses of Indiana*, fig. 61.

Upper Peninsula: Alger and Keweenaw counties, *Parmelee*. Mackinac Co., *Wheeler*. Marquette Co., *Nichols*. Ontonagon Co., *Darlington*.

Lower Peninsula: Charlevoix and Cheboygan counties, *Wheeler*. Emmet Co., *Parmelee*. Ingham Co., *Wheeler*.

8. *Dicranodontium* BSG

Leaves narrow, long-subulate, channeled, serrulate; costa broad, filling the upper part of the leaf; alar cells inflated; dioicous; setae curved to cygneous when moist; capsules erect and symmetric, smooth; peristome teeth split nearly to the base.

[57]

Dicranodontium denudatum (Brid.) E. Britt. — Leaves erect-spreading or secund, setaceous; costa 1/3 – 1/2 the width of the leaf base; spores 10–15 μ, maturing in late autumn or winter. — On moist, shaded rock or other substrata. Widespread in eastern North America; Alaska and British Columbia; Europe and Japan.

ILLUSTRATIONS: Figure 45. Grout, *Moss Flora of North America*, vol. 1, Pl. 43C (as *D. longirostre*).

Upper Peninsula: Alger and Luce counties, *Nichols*.

LEUCOBRYACEAE

Grayish or whitish-green plants in dense cushions; stems erect; leaves thick and fleshy, consisting almost entirely of costa of 2 or more layers of large, empty, hyaline cells enclosing a layer of small, green cells; setae terminal, mostly elongate; capsules erect and symmetric or inclined and asymmetric; peristome single, consisting of 8 or 16 entire or bifid teeth.

Leucobryum Hampe

Plants in conspicuous, spongy, whitish cushions; leaves erect-spreading or secund, lanceolate, subtubulose above; costa very broad, filling most of the leaf; capsule curved, striate, strumose.

Leucobryum glaucum (Hedw.) Schimp. — Leaves crowded, oblong-lanceolate from an ovate-lanceolate base; setae 2–3 cm. long; peristome bifid to the middle; spores in autumn. — On soil or humus in open woods and at the margins of swamps. Widespread in eastern North America; Europe, Macaronesia, Caucasus, and Japan.

ILLUSTRATIONS: Figure 46. Conard, *How to Know the Mosses* (ed. 2), fig. 29. Grout, *Mosses with Hand-Lens and Microscope*, fig. 49. Jennings, *Mosses of Western Pennsylvania* (ed. 2), Pl. 14. Welch, *Mosses of Indiana*, fig. 62.

Upper Peninsula: Alger Co., *Steere*. Chippewa Co., *Thorpe*. Gogebic Co., *Darlington*. Marquette Co., *Nichols*. Ontonagon Co., *Bessey*.
Lower Peninsula: Cheboygan Co., *Nichols*. Clinton Co., *Marshall*. Crawford Co., *Parmelee*. Eaton Co., *Darlington*. Emmet Co., *Wheeler*. Ingham and Ionia counties, *Parmelee*. Leelanau Co., *Darlington*. Macomb Co., *Cooley*. Oakland Co., *Darlington*. Oscoda Co., *Parmelee*. Washtenaw Co., *Kauffman*.

ENCALYPTACEAE

Stems erect; leaves broad, mostly lingulate, often awned; costa single, well developed; upper cells short, densely papillose (or rarely strongly mammillose); lower cells pale, smooth, rectangular, with thin side- and thickened end-walls; alar cells not differentiated;

[58]

setae elongate; capsules erect and symmetric, cylindric; peristome lacking, single, or double; operculum long-rostrate; calyptra large, long-campanulate, generally fringed at base.

Encalypta Hedw.

Plants small to medium-sized in dull tufts on sandy or gravelly soil or crevices of rocks (usually calcareous); stems simple or forked, densely foliate; leaves mostly lingulate, usually folded and incurved at the tips when dry; upper cells papillose; capsules cylindric, smooth or plicate.

1. Leaves blunt or rounded at apex (or only the uppermost leaves of fertile stems awned); filiform brood-bodies in leaf axils of sterile stems; spores 8–18 μ 2
1. Leaves usually short-pointed or awned; brood-bodies absent; spores 30 μ or more 3
2. Stems without a central strand; leaves not awned, strongly crisped when dry . . . . . . 3. *E. streptocarpa*
2. Stems with a small, indistinct central strand; upper leaves awned, only slightly crisped . . . . . 2 *E. procera*
3. Leaves hyaline-tipped or awned; capsules furrowed; spores 35–55 μ, warty . . . . 4. *E. rhaptocarpa*
3. Leaves usually abruptly short-pointed; capsules smooth; spores 31–36 μ, irregularly reticulate . . . . . 1. *E. ciliata*

1. *Encalypta ciliata* Hedw. — Leaves oblong-ovate to lingulate, imbricate and crisped when dry; costa stout, reddish, subpercurrent or shortly excurrent; capsules smooth, 2–3 mm. long; calyptra strongly fringed. — On shaded soil, especially in rock crevices. Across Canada, Alaska and the northern United States; Europe, Africa and Asia.

ILLUSTRATIONS: Figure 47. Conard, *How to Know the Mosses* (ed. 2), fig. 77. Grout, *Mosses with Hand-Lens and Microscope*, Pl. 37.

Upper Peninsula: Alger Co., *Steere.* Baraga Co., *Parmelee.* Keweenaw Co., *Steere.* Marquette Co., *Nichols.* Ontonagon Co., *Nichols & Steere.*
Lower Peninsula: Alpena Co., *Robinson & Wells.* Cheboygan Co., *Nichols.* Eaton Co., *Wheeler.*

2. *Encalypta procera* Bruch — Sterile stems often bearing filamentous brood-bodies; central strand none; leaves not much crisped, lingulate, obtuse, often awned; capsules spirally furrowed; peristome double. — On soil or rocks. Canada and Alaska, south to the Great Lakes; Europe and Asia.

ILLUSTRATION: Grout, *Moss Flora of North America*, vol. 1, Pl. 72.

Upper Peninsula: Keweenaw Co., *Hermann.* Marquette Co., *Hill.*
Lower Peninsula: Alpena Co., *Robinson & Wells.*

3. *Encalypta streptocarpa* Schwaegr. — Closely related to *E. pro-cera* but sterile in North America; stems bearing filamentous brood-bodies in leaf axils; central strand present (but small and indistinct); leaves crisped, not awned. — On soil and rock. Across the continent, south to Iowa and the Great Lakes; Europe, Canary Islands, and Asia.

ILLUSTRATIONS: Conard, *How to Know the Mosses* (ed. 2), fig. 76a–d. Grout, *Mosses with Hand-Lens and Microscope*, fig. 80. Jennings, *Mosses of Western Pennsylvania* (ed. 2), Pl. 62.

Upper Peninsula: Keweenaw and Marquette counties, *Steere*. Ontonagon Co., *Nichols & Steere*.
Lower Peninsula: Alpena Co., *Robinson & Wells*. Cheboygan Co., *Nichols*. Emmet Co., *Steere*. Leelanau Co., *Darlington*. Mackinac Co., *Nichols*. Washtenaw Co., *Steere*.

4. *Encalypta rhaptocarpa* Schwaegr. — Leaves ± awned; capsules straight-ribbed; peristome single; spores 35–55 μ, warty. — On damp soil and rock. Across the northern United States, Canada, and Alaska; Europe, Asia and Hawaii.

ILLUSTRATIONS: Conard, *How to Know the Mosses* (ed. 2), fig. 76e–f. Grout, *Moss Flora of North America*, vol. 1, Pl. 69B.

Lower Peninsula: Cheboygan Co., *Nichols*.

POTTIACEAE

Plants very small to large. Stems erect, usually tufted, radiculose at base; leaves various in shape, mostly entire; costa single, well developed; upper cells usually small and thick-walled, usually ± papillose; lower cells rectangular, thin-walled and hyaline; alar cells not differentiated; setae mostly elongate; capsules mostly erect and symmetric; operculum usually differentiated; peristome single, the teeth sometimes deeply cleft into filiform, often twisted segments (or rarely lacking).

1. Plants small or minute (less than 1 cm. high); capsules
   (except in *Weissia*) dehiscing irregularly 2
1. Plants larger; capsules operculate 5
   2. Upper leaf margins strongly involute 3
   2. Upper leaf margins ± revolute 4
3. Capsules exserted, oblong-cylindric; spores 15–18 μ . . 2. *Weissia*
3. Capsules immersed, elongate-spherical; spores 20–30 μ . 1. *Astomum*
   4. Capsules mostly apiculate; leaf cells papillose . 12. *Phascum*
   4. Capsules not apiculate; leaf cells mostly smooth . 11. *Acaulon*
5. Peristome none 6
5. Peristome of 16 teeth 10

[60]

6. Operculum remaining attached to the columella after dehiscence; leaf margins recurved on 1 side below . . . . 5. *Hymenostylium*
6. Operculum not attached to the columella after dehiscence; leaf margins various, not recurved on 1 side 7
7. Leaves broad, with margins strongly inrolled when dry 8. *Hyophila*
7. Leaves narrow or broad, not inrolled at the margins when dry . . 8
8. Costa typically excurrent; spores 23–30 μ . . . 13. *Pottia*
8. Costa vanishing below the apex; spores 8–12 μ 9
9. Annulus none; capsules ovoid to short-oblong . 3. *Gymnostomum*
9. Annulus large, persistent; capsules narrowly oblong . 4. *Gyroweisia*
10. Leaves elliptic or lingulate, ± involute at the margins especially when dry . . . . . 14. *Desmatodon*
10. Leaves narrow, mostly lanceolate, revolute at the margins 11
11. Leaves broad, widest above the base, mostly rounded at the apex, generally awned . . . . 15. *Tortula*
11. Leaves narrow, widest near the base, acute, sometimes subulate but not awned 12
12. Leaf margins ± revolute 13
12. Leaf margins plane or incurved 14
13. Peristome teeth erect . . . . 10. *Didymodon*
13. Peristome teeth spirally twisted . . . . . . 9. *Barbula*
14. Basal cells extending beyond the leaf shoulders as a short, hyaline border . . . . 6. *Tortella*
14. Basal cells not extending beyond the shoulders . 7. *Trichostomum*

1. *Astomum* Hampe

Plants minute, growing on soil in open places; leaves slender, with involute margins; capsules immersed; operculum usually differentiated but non-functional; peristome none.

Astomum muehlenbergianum (Sw.) Grout — Small, densely tufted plants resembling sterile *Weissia controversa;* leaves strongly crisped when dry, somewhat sheathing at base; margins involute above, entire; costa excurrent as a strong mucro; upper cells quadrate to hexagonal, pluripapillose; capsules elongate-spherical, apiculate; spores 20–30 μ, strongly papillose, mature in early spring. — On soil in open fields. Massachusetts to Saskatchewan, south to Alabama and Texas; Arizona; eastern Asia.

ILLUSTRATIONS: Figure 48. Conard, *How to Know the Mosses* (ed. 2), fig. 75a–b. Jennings, *Mosses of Western Pennsylvania* (ed. 2), Pl. 62.

Lower Peninsula: Kalamazoo Co., *Becker.* Washtenaw Co., *Steere.*

2. *Weissia* Hedw.

Small, caespitose plants, usually growing on bare, dry soil; leaves narrow, involute at margins above; costa well developed; cells small,

[61]

densely papillose; capsules exserted, erect and symmetric, ovoid to cylindric, operculate; peristome often poorly developed.

Weissia controversa Hedw. — Plants yellow-green, 3–5 mm. tall; leaves lanceolate or linear-lanceolate, acute and mucronate, with strongly involute upper margins; costa shortly excurrent; capsules dark-brown, lustrous; operculum long-rostrate; spores 15–20 μ, mature in spring. — A weed, generally found on dry soil in open places. Widely distributed in North America; almost cosmopolitan.

ILLUSTRATIONS: Figure 49. Conard, *How to Know the Mosses* (ed. 2), fig. 84. Grout, *Mosses with Hand-Lens and Microscope*, Pl. 23. Jennings, *Mosses of Western Pennsylvania* (ed. 2), Pl. 16. Welch, *Mosses of Indiana*, fig. 67.

Upper Peninsula: Alger Co., *Parmelee.*
Lower Peninsula: Cheboygan Co., *Nichols.* Eaton, Ingham, and Oakland counties, *Darlington.* Washtenaw Co., *Steere.*

3. Gymnostomum Nees, Hornsch., & Sturm

Densely tufted plants usually growing on moist to dripping limestone ledges or cliffs; stems erect, forked, ± rusty-brown below; leaves linear-lanceolate, densely papillose; costa well developed; setae elongate; capsules erect, ovoid to oblong, glossy when mature; operculum long-rostrate; peristome none.

Gymnostomum aeruginosum Sm. — Plants yellow- or olive-green; leaves crowded, larger above, acute or subacute, often blunt at the tip; costa stout; cells small, thick-walled, densely papillose and obscure; capsules yellow-brown; spores 10–12 μ, mature in late summer or autumn. — On moist, calcareous rocks. Eastern North America; Arizona; Europe, Canary Islands, and Asia.

Especially blunt-leaved forms of this species are frequently referred to *G. calcareum*, which is known from only a few North American localities (in California).

ILLUSTRATIONS: Figure 50. Conard, *How to Know the Mosses* (ed. 2), fig. 97d–e. Grout, *Mosses with Hand-Lens and Microscope*, Pl. 25 (as *G. rupestre*).

Upper Peninsula: Alger Co., *Steere.* Keweenaw Co., *Hermann.* Mackinac and Marquette counties, *Nichols.*
Lower Peninsula: Eaton Co., *Becker.*

4. Gyroweisia (Schimp.) Schimp.

Small, rock-inhabiting mosses closely related to *Gymnostomum*, but differing in stems with a central strand, sheathing perichaetial leaves, and a large, persistent annulus.

Gyroweisia tenuis (Hedw.) Schimp. — Plants 1–2 mm. high, in thin tufts; leaves narrowly ligulate, rounded-obtuse, plane-margined; costa ending at or below the apex; upper cells irregularly sub-quadrate (sometimes 2–3:1), densely papillose; capsules oblong; 8–10 μ, maturing in late summer. — Rare and local, on rock. Manitoba, Michigan, and Iowa; Europe, North Africa, Madeira, and the Middle East.

ILLUSTRATIONS: Figure 51. Grout, *Moss Flora of North America*, vol. 1, Pl. 74C.

Upper Peninsula: Keweenaw Co., *Steere*. Houghton Co., *Steere & Sharp*.

5. *Hymenostylium* Brid.

Plants in dense tufts on wet, calcareous rocks; leaves lanceolate, keeled, acute; margins narrowly recurved on 1 side below; costa strong, mostly ending just below the apex; cells densely papillose; dioicous; setae elongate; capsules erect and symmetric; operculum remaining attached to the columella for some time after dehiscence and finally falling with it; peristome none.

Hymenostylium recurvirostrum (Hedw.) Dix. — Tufts 2–5 cm. high, green above, rusty-brown below; leaves lance-acuminate, acute, 1–1.5 mm. long; costa vanishing somewhat below the apex; cells short, rounded-quadrate or short-rectangular; setae 8–10 mm. long; capsules red-brown, ovoid or rounded-ovoid; spores about 15 μ, maturing in summer. — On moist or wet, calcareous cliffs. Greenland; Canada southward to the Carolinas and California; Europe, Asia, and Africa.

ILLUSTRATIONS: Figure 52. Conard, *How to Know the Mosses* (ed. 2), fig. 97a–c (as *Gymnostomum*). Grout, *Mosses with Hand-Lens and Microscope*, Pl. 24 (as *Gymnostomum*). Jennings, *Mosses of Western Pennsylvania* (ed. 2), Pl. 16. Welch, *Mosses of Indiana*, fig. 69 (as *Gymnostomum*).

Upper Peninsula: Alger Co., *Wheeler*. Luce and Marquette counties, *Nichols*. Ontonagon Co., *Nichols & Steere*.
Lower Peninsula: Alpena Co., *Robinson & Wells*. Eaton Co., *Wheeler*.

6. *Tortella* (C.M.) Limpr.

Leaves oblong- to linear-lanceolate, acute to subulate, mostly crisped, often curled when dry; margins erect and ± wavy; costa well developed, ending in the apex or excurrent; upper cells small, densely papillose, abruptly differentiated from the larger, hyaline basal cells which extend beyond the shoulders to form a V-shaped area; setae elongate; capsules ellipsoidal to cylindric, erect and symmetric; operculum rostrate; peristome of 32 filiform, ± twisted, papillose divisions.

1. Leaves little crisped when dry, very shiny below, abruptly
   tapered to a long, fragile tip consisting almost
   wholly of costa . . . . 3. *T. fragilis*
1. Leaves ± crisped when dry, not especially shiny
   below, not fragile or abruptly subulate 2
   2. Leaf margins incurved and apex cucullate . . 4. *T. inclinata*
   2. Leaf margins plane or slightly incurved above 3
3. Leaves broadly acute and abruptly mucronate; stems with
   a central strand; plants monoicous . . . . . 1. *T. humilis*
3. Leaves acuminate; central strand lacking;
   plants dioicous . . . . 2. *T. tortuosa*

1. *Tortella humilis* (Hedw.) Jenn.

1. *Tortella humilis* (Hedw.) Jenn. — Tufts dense, green or yellow-green, up to 3 cm. high; leaves oblong-lanceolate, rather broadly acute and mucronate; costa shortly excurrent; setae about 1.5 cm. long; capsules oblong-cylindric; spores 7–10 μ, maturing in spring. — On soil or rotten wood. Widespread in eastern North America; Europe, North Africa and the Caucasus.

ILLUSTRATIONS: Conard, *How to Know the Mosses* (ed. 2), fig. 94a–c. Grout, *Mosses with Hand-Lens and Microscope*, Pl. 32 (as *T. caespitosa*). Jennings, *Mosses of Western Pennsylvania* (ed. 2), Pl. 17. Welch, *Mosses of Indiana*, fig. 71.

Upper Peninsula: Ontonagon Co., *Nichols & Steere.*

Lower Peninsula: Benzie Co., *Darlington.* Cheboygan Co., *Nichols.* Clinton Co., *Parmelee.* Leelanau Co., *Darlington.* Washtenaw Co., *Steere.*

2. *Tortella tortuosa* (Hedw.) Limpr.

2. *Tortella tortuosa* (Hedw.) Limpr. — Tufts dull, yellowish or brownish; leaves 2–6.5 mm. long, strongly twisted and spirally curled when dry, oblong-lanceolate, slenderly long-acuminate; costa excurrent; setae 1.7–3 cm. long, reddish below, paler above; spores maturing in late spring or in summer. — On sandy or gravelly soil or in rocks. Greenland; across the continent, south to the Gulf of Mexico; Europe, Asia, North Africa, and the Canary Islands.

ILLUSTRATIONS: Figure 53. Conard, *How to Know the Mosses* (ed. 2), fig. 94d. Grout, *Mosses with Hand-Lens and Microscope*, fig. 73.

Upper Peninsula: Dickinson Co., *Wheeler.* Houghton Co., *Hyypio.* Keweenaw Co., *Parmelee.* Marquette Co., *Nichols.* Ontonagon Co., *Darlington.*

Lower Peninsula: Alpena Co., *Robinson & Wells.* Cheboygan and Emmet counties, *Nichols.* Huron Co., *Schnooberger.* Leelanau Co., *Darlington.*

3. *Tortella fragilis* (Hook.) Limpr.

3. *Tortella fragilis* (Hook.) Limpr. — Tufts green or brown, rigid; leaves not much contorted when dry, rapidly narrowed from a shiny, lanceolate base to a long, slender, fragile subula consisting almost entirely of excurrent costa; upper cells 7–11 μ, densely papillose; capsules 2–3 mm. long; spores in summer (but seldom

[64]

fruiting). — On sand and rocks, particularly along lake shores. Across the continent, south to New Jersey and Missouri; Europe, North Africa, and Asia.

ILLUSTRATION: Grout, *Mosses with Hand-Lens and Microscope*, Pl. 31.

Upper Peninsula: Alger Co., *Hermann.* Keweenaw Co., *Steere.* Ontonagon Co., *Nichols & Steere.*
Lower Peninsula: Alpena Co., *Robinson & Wells.* Cheboygan Co., *Nichols.* Huron Co., *Schnooberger.*

4. *Tortella inclinata* Hedw. f. — Tufts brown; leaves crisped when dry, oblong-lanceolate, obtuse and ± cucullate at apex; costa short-excurrent as a mucro; upper cells 10–13 μ wide; perichaetial leaves distinctly differentiated, long-linear, mostly hyaline throughout; setae 1–2 cm. long; capsules narrowly cylindric. — On gravel and sand, mostly in open places. Southern Canada and the Great Lakes region of the United States; Europe, Caucasus, and the Middle East.

ILLUSTRATION: Grout, *Moss Flora of North America*, vol. 1, Pl. 84 (as *T. inclinatula*).

Lower Peninsula: Cheboygan and Emmet counties, *Sharp.*

7. *Trichostomum* Bruch

Leaves oblong-lanceolate or lingulate, ± concave; margins plane or incurved, often wavy; costa percurrent or shortly excurrent; upper cells small, rounded-hexagonal, densely papillose; lower cells pale, oblong, thin-walled, smooth, not extending up the margins at the shoulders; setae elongate; capsules erect and symmetric, ± cylindric; operculum rostrate; peristome teeth erect, entire, perforate, or deeply cleft.

Trichostomum cylindricum (Bruch) C.M. — Leaves narrowly oblong-lanceolate, gradually acuminate, crisped when dry, slightly wavy at the margins; costa percurrent or short-excurrent; upper cells about 7–10 (rarely 17) μ; dioicous; spores 12–14 μ. — On damp, shaded rocks. Greenland; widespread in eastern North America; Arizona; Europe, Africa, Asia, and South America.

ILLUSTRATIONS: Figure 54. Grout, *Mosses with Hand-Lens and Microscope*, Pl. 30.

Upper Peninsula: Keweenaw Co. (Fort Wilkins), *Hermann.*

8. *Hyophila* Brid.

Dark-green, loosely tufted plants; leaves crowded, ± spatulate, with margins inrolled when dry; costa mostly ending somewhat be-

[65]

low the apex; upper cells small, rounded-hexagonal, usually papillose; lower cells oblong, hyaline; setae elongate; capsules erect and symmetric, smooth, cylindric to ovoid; peristome absent.

Hyophila tortula (Schwaegr.) Hampe — Plants up to 3 cm. high (but usually less); leaves lingulate, broadly pointed, usually irregularly serrate above; dioicous; capsules cylindric, about 1 mm. long; operculum conic to rostrate; annulus deciduous; peristome none; brood-bodies on branched stalks often produced in leaf axils. — On wet or moist rocks, frequently in stream beds. Widespread in eastern North America and throughout tropical America; Europe.

ILLUSTRATION: Figure 55.

Upper Peninsula: Alger and Luce counties, *Sharp.*

9. *Barbula* Hedw.

Small to medium-sized, dull, green or brownish, caespitose plants on soil or rocks; leaves oblong to linear-lanceolate, not awned, mostly contorted when dry; margins generally recurved; costa strong, ending near the apex to shortly excurrent; cells small, obscure, mostly papillose; setae elongate; capsules erect, oblong to cylindric; operculum long-rostrate; peristome teeth 16, deeply divided into 32 filiform, twisted segments.

1. Stems red, with numerous axillary propagula . 1. *B. michiganensis*
1. Stems yellow or yellowish-brown, without propagula 2
   2. Setae yellow; perichaetial leaves convolute . . 2. *B. convoluta*
   2. Setae red; perichaetial leaves not convolute 3
3. Leaves blunt and mucronate . . . . . . 3. *B. unguiculata*
3. Leaves acute or subacute, not mucronate 4
   4. Leaves usually less than 2.5 mm. long;
     setae 1–1.5 cm. long . . . . 4. *B. fallax*
   4. Leaves usually more than 2.5 mm. long; setae
     1.5–3 cm. long . . . . 5. *B. cylindrica*

1. *Barbula michiganensis* Steere — Small plants in dense, yellow-green tufts; stems red; propagula of 2–6 cells produced in leaf axils; leaves narrowed from a broadly ovate, clasping base; cells rounded to oval, thick-walled. — On soft sandstone cliffs; endemic.

ILLUSTRATION: Grout, *Moss Flora of North America*, vol. 1, Pl. 89.

Upper Peninsula: Alger Co. (Pictured Rocks), *Nichols & Steere.*

2. *Barbula convoluta* Hedw. — Small, densely tufted, yellowish or glaucous-green plants; leaves oblong, obtuse, mucronate, incurved-crisped when dry; costa ending near the apex; perichaetial leaves

conspicuously convolute; setae yellow; spores in spring. — On sandy or gravelly soil in open places. Widespread in North America, North Africa, Asia, and Macaronesia.

ILLUSTRATIONS: Figure 56. Grout, *Mosses with Hand-Lens and Microscope,* Pl. 28. Jennings, *Mosses of Western Pennsylvania* (ed. 2), Pl. 17.

Upper Peninsula: Keweenaw Co., *Hermann.* Luce Co., *Nichols.*
Lower Peninsula: Alpena Co., *Robinson & Wells.* Cheboygan Co., *Nichols.* Ingham Co., *Parmelee.* Washtenaw Co., *Steere.*

3. *Barbula unguiculata* Hedw. — Plants bright-green, becoming dull-green or brown; leaves contorted when dry, oblong-lanceolate, mucronate or cuspidate; costa shortly excurrent; perichaetial leaves not much differentiated; setae red; capsules oblong-cylindric, chestnut-brown; spores in winter or early spring. — Common on soil of roadside banks and fields. Newfoundland to North Carolina, west to Montana; Europe, North Africa, Azores, and Asia.

ILLUSTRATIONS: Conard, *How to Know the Mosses* (ed. 2), fig. 92. Grout, *Mosses with Hand-Lens and Microscope,* fig. 72. Jennings, *Mosses of Western Pennsylvania* (ed. 2), Pl. 17. Welch, *Mosses of Indiana,* fig. 73.

Upper Peninsula: Mackinac Co., *Steere.*
Lower Peninsula: Alpena Co., *Robinson & Wells.* Cheboygan Co., *Steere.* Eaton Co., *Darlington.* Ingham Co. *Parmelee.* Kalamazoo Co., *Becker.* Leelanau and Oakland counties, *Darlington.* Washtenaw Co., *Kauffman.*

4. *Barbula fallax* Hedw. — Plants dull-green or brownish, 1–3 cm. high; leaves lance-acuminate from an ovate base, keeled; costa ending in the apex or slightly excurrent; cells small, obscure, somewhat papillose; setae 1–1.5 cm. long; capsules long-ovoid to subcylindric; peristome bright-red; spores smooth, in late summer to early spring. — On shaded banks and ledges. Greenland; Nova Scotia to Virginia, Iowa and Montana; Europe, North Africa, Madeira, and Asia.

ILLUSTRATIONS: Conard, *How to Know the Mosses* (ed. 2), fig. 92a–b. Grout, *Mosses with Hand-Lens and Microscope,* Pl. 29. Welch, *Mosses of Indiana,* fig. 74.

Upper Peninsula: Alger Co., *Steere.* Gogebic Co., *Conard.* Keweenaw Co., *Hermann.* Mackinac Co., *Steere.*
Lower Peninsula: Alpena Co., *Robinson & Wells.* Cheboygan Co., *Steere.* Eaton, Ingham, and Leelanau counties, *Darlington.* Presque Isle and Washtenaw counties, *Steere.*

5. *Barbula cylindrica* (Tayl.) Schimp. — Plants dark-green, 1–6 cm. high; upper leaves larger, lance-subulate, contorted when dry; costa disappearing in the apex; upper cells small, obscure, very papillose, irregularly quadrate; setae 1.5–3 cm. long; spores in late spring. — On soil. Michigan; Alaska to Mexico; Europe, North Africa, and Asia.

[67]

ILLUSTRATION: Grout, *Moss Flora of North America*, vol. 1, Pl. 86.

Upper Peninsula: Keweenaw Co., *Steere.*

10. *Didymodon* Hedw.

Plants similar in most respects to *Barbula*, caespitose on soil or rock; leaves mostly narrow, with margins generally revolute; costa strong; upper cells small, rounded-quadrate, usually papillose; setae elongate; capsules oblong to cylindric, mostly erect; annulus sometimes differentiated; operculum conic-rostrate; peristome teeth erect, sometimes deeply divided.

1. Leaves ovate-lanceolate . . . . 3. *D. trifarius*
1. Leaves narrower 2
  2. Plants rusty-red below; leaves irregularly
    dentate near the apex . . . . 1. *D. recurvirostris*
  2. Plants brownish or brownish-green below; leaves entire 3
3. Leaves decurrent, narrowed above to a strap-shaped,
  rounded-obtuse apex; costa ending below the apex; leaf
  margins unistratose . . . . 4. *D. tophaceus*
3. Leaves not decurrent, acuminate; costa disappearing
  in the apex; leaf margins bistratose above . . . 2. *D. rigidulus*

1. *Didymodon recurvirostris* (Hedw.) Jenn. — Plants green above, brick-red below; leaves lanceolate from a broader, white or reddish base, irregularly dentate above; costa usually shortly excurrent as a pellucid mucro; basal cells lax and thin-walled, pale, oblong; capsules erect; annulus revoluble; spores in summer or early autumn. — On various substrata, soil, rock, or logs, often in wet, calcareous habitats. Greenland; across Canada, Alaska and the northern United States; Europe, Africa, Asia, and New Zealand.

ILLUSTRATIONS: Figure 57. Conard, *How to Know the Mosses* (ed. 2), fig. 91a–d. Grout, *Mosses with Hand-Lens and Microscope*, Pl. 27 (as *D. rubellus*).

Upper Peninsula: Keweenaw Co., *Steere.* Mackinac Co., *Nichols.* Ontonagon Co., *Nichols* & *Steere.*

Lower Peninsula: Alpena Co., *Robinson* & *Wells.* Cheboygan and Emmet counties, *Steere.* Kalamazoo Co., *Becker.* Leelanau Co., *Darlington.* Livingston and Presque Isle counties, *Steere.*

2. *Didymodon rigidulus* Hedw. — Leaves lance-subulate, with margins recurved below, bistratose above; stalked, subspherical propagula in leaf axils. — On calcareous soil or rocks. Great Lakes region of Canada and the United States, west to California and Arizona; Europe, Azores, and Asia.

ILLUSTRATION: Grout, *Moss Flora of North America*, vol. 1, Pl. 90B.

Upper Peninsula: Delta and Marquette counties, *Steere.*
Lower Peninsula: Alpena Co., *Robinson* & *Wells.*

3. *Didymodon trifarius* (Hedw.) Brid. — Plants densely tufted, olive-green; leaves broad, ovate-lanceolate, obtuse or acute; costa ending in or below the apex; cells small and rounded, incrassate, smooth; capsules oblong to short-cylindric; operculum conic, short-rostrate; spores in winter. — On calcareous soil and rock. New York to British Columbia, more common in the West; Europe, Macaronesia, North Africa, and the Middle East.

ILLUSTRATIONS: Conard, *How to Know the Mosses* (ed. 2), fig. 91e. Grout, *Mosses with Hand-Lens and Microscope*, fig. 70 (as *D. luridus*). Welch, *Mosses of Indiana*, fig. 75.

Upper Peninsula: Mackinac Co. (Mackinac I.), *Nichols.*

4. *Didymodon tophaceus* (Brid.) Lisa — Plants often embedded in calcareous tufa; leaves erect-spreading when moist, incurved and not much contorted when dry, lanceolate or lingulate from a broader, decurrent base, obtuse or rarely ± acute; costa ending below the apex; cells irregularly rounded, slightly papillose, incrassate; capsules elliptic to long-cylindric; spores in winter. — On wet, calcareous soil or rock. New York to British Columbia, south to Tennessee and Arizona; Mexico; Europe, Africa, and Asia.

ILLUSTRATIONS: Grout, *Mosses with Hand-Lens and Microscope*, fig. 71. Welch, *Mosses of Indiana*, fig. 76.

Lower Peninsula: Eaton Co., *Steere.* Kalamazoo Co., *Becker.*

11. *Acaulon* C.M.

Plants minute and bud-like; leaves ovate, apiculate, concave; costa thin, ending in the apex or shortly excurrent; upper cells quadrate to rectangular or shortly rhomboidal, incrassate, mostly smooth or nearly so; capsules immersed, subglobose, not apiculate, irregularly dehiscent.

Plants ± 3-cornered; costa excurrent; spores rough . 1. *A. triquetrum*
Plants subglobose; costa usually not excurrent;
spores smooth . . . . 2. *A. rufescens*

1. *Acaulon triquetrum* (Spruce) C.M. — Plants up to 1 mm. high, bulbiform, ± 3-cornered; costa excurrent; spores up to 30 μ, minutely spinose, maturing in early spring. — On soil. New England to western Canada, south to South Carolina; Europe.

ILLUSTRATIONS: Figure 58. Grout, *Mosses with Hand-Lens and Microscope*, fig. 67.

Lower Peninsula: Clinton Co. (Rose Lake Experiment Station), *Marshall* (also Jackson, Kalamazoo, and Washtenaw counties, according to *Steere*).

2. *Acaulon rufescens* Jaeg. & Sauerb. — Plants subglobose, not 3-cornered; leaf margins sharply reflexed and often coarsely dentate; costa ending slightly below the apex to (rarely) short-excurrent; spores 40–50 μ, smooth. — On clay or sand. Widespread east of the Rocky Mountains; Arizona.

ILLUSTRATIONS: Grout, *Moss Flora of North America*, vol. 1, Pl. 91, 92. Welch, *Mosses of Indiana*, fig. 77.

Lower Peninsula: Kalamazoo Co., *Becker*. Washtenaw Co., *Steere*.

12. *Phascum* Hedw.

Plants small or very small, growing on soil in open situations; stems simple or sometimes branched; upper leaves largest, lanceolate to ovate-lanceolate; costa excurrent; upper cells quadrate or hexagonal, ± papillose; capsules immersed or emergent, subglobose to ovoid, apiculate, dehiscing irregularly.

Phascum cuspidatum Hedw. — Plants densely caespitose, green; leaves oblong-lanceolate, entire; costa excurrent as a cuspidate point; capsules ovoid-globose, nearly or quite immersed; spores 24–35 μ, maturing in spring. — On sand or clay of banks, old fields, roadsides, etc. Widespread in North America but rare or overlooked; Europe, North Africa, and the Caucasus.

ILLUSTRATIONS: Figure 59. Grout, *Mosses with Hand-Lens and Microscope*, fig. 68. Welch, *Mosses of Indiana*, fig. 78.

Lower Peninsula: Clinton Co., *Marshall*. Kalamazoo Co., *Becker*. Washtenaw Co., *Steere*.

13. *Pottia* Fürnr.

Small plants on bare, calcareous soil, gregarious to caespitose; leaves broadly lanceolate or ovate, often keeled; costa ending in the apex to excurrent; upper cells quadrate to hexagonal or rhombic, smooth or ± papillose; capsules exserted, erect and symmetric; peristome present or lacking; operculum conic.

Pottia truncata (Hedw.) Fürnr. — Plants 3–5 mm. high, usually gregarious; leaves lanceolate to spatulate, acute; margins plane and slightly crenulate above, narrowly recurved below; costa excurrent as a stout, smooth awn; upper cells smooth or nearly so; capsules 0.4–0.8 mm. long, top-shaped; peristome none; spores in late spring or early summer. — On moist soil in open situations (fields, meadows, roadsides). Nova Scotia to the Great Lakes and Maryland; Europe, Asia, and Macaronesia.

ILLUSTRATIONS: Figure 60. Grout, *Mosses with Hand-Lens and Microscope*, Pl. 33 (as *P. truncatula*). Jennings, *Mosses of Western Pennsylvania* (ed. 2), Pl. 17.

Lower Peninsula: Eaton Co., *Parmelee*. Kalamazoo Co., *Becker*. Washtenaw Co., *Steere*.

14. *Desmatodon* Brid.

Plants mostly densely tufted, on soil or rocks; leaves oblong-lanceolate to ovate or obovate, sometimes bordered by short, pale cells or by linear cells; costa well developed, mostly excurrent as a point or awn; cells short, smooth or papillose; setae elongate, usually twisted; capsules mostly erect, oblong to cylindric; peristome teeth 16, divided nearly to the base, not or only slightly twisted.

Desmatodon obtusifolius (Schwaegr.) Schimp. — Leaves oblong-lingulate, broadly acute to rounded-obtuse, often mucronate; costa ending somewhat below the apex to slightly excurrent; upper cells subquadrate, obscure, densely papillose; setae about 1 cm. long; capsules erect, 2–4 mm. long; peristome teeth short, straight. — Occasional on moist calcareous or siliceous rocks. New Brunswick to British Columbia, south to Pennsylvania, Missouri, and Arizona; Europe, Asia, and North Africa.

ILLUSTRATIONS: Figure 61. Conard, *How to Know the Mosses* (ed. 2), fig. 82a–c. Jennings, *Mosses of Western Pennsylvania* (ed. 2), Pl. 17.

Upper Peninsula: Alger Co., *Steere*. Marquette Co., *Nichols*.
Lower Peninsula: Eaton Co., *Steere*. Huron Co., *Schnooberger*.

15. *Tortula* Hedw.

Plants small to large, generally tufted, growing on soil and rocks or sometimes on bark; leaves mostly oblong or obovate, often rounded at apex and mostly awned, with margins ± revolute; costa mostly excurrent; upper cells small, ± hexagonal, densely papillose; lower cells enlarged, oblong, pale, smooth; setae elongate; capsules erect and symmetric, cylindric; operculum rostrate; peristome teeth divided to a well-developed basal membrane into 32 long, filiform segments which are twisted together when dry.

1. Very small plants on bark of trees; leaves bearing
   propagula . . . . 1. *T. papillosa*
1. Large plants on soil or rock; leaves not bearing
   propagula 2
   2. Leaves acute or acuminate, ending in a short, concolorous point
      or a short awn; upper cells weakly papillose . 2. *T. mucronifolia*
   2. Leaves blunt to emarginate, ending in a long, white or reddish hair-
      point; upper cells densely papillose . . . . 3. *T. ruralis*

[71]

1. *Tortula papillosa* Wils. — Small, dark-green plants, scattered or in small tufts; leaves obovate, rounded to emarginate at apex; costa excurrent as a mucro or short awn; propagula frequently produced on the leaves. — On tree trunks. Ontario; New England to Illinois and North Carolina; Europe; Mexico and Colombia; Australia.

ILLUSTRATIONS: Grout, *Mosses with Hand-Lens and Microscope*, fig. 79. Jennings, *Mosses of Western Pennsylvania* (ed. 2), Pl. 62.

Lower Peninsula: Allegan Co., *Schnooberger*. Clinton Co., *Parmelee*. Gratiot and Isabella counties, *Schnooberger*. Livingston Co., *Steere*. Montcalm, Ottawa, and Van Buren counties, *Schnooberger*. Washtenaw Co., *Steere*.

2. *Tortula mucronifolia* Schwaegr. — Plants bluish- or dark-green, in rather soft, often dense tufts; leaves long-spatulate, acute or acuminate; costa excurrent as a yellow mucro or spine; cells weakly papillose; spores in late summer, 14–18 μ. — On moist soil or rock. Greenland; across the continent, south to New York, Iowa, and New Mexico; Europe and Asia.

ILLUSTRATIONS: Conard, *How to Know the Mosses* (ed. 2), fig. 146. Grout, *Mosses with Hand-Lens and Microscope*, Pl. 36. Welch, *Mosses of Indiana*, fig. 81.

Upper Peninsula: Keweenaw Co., *Parmelee*. Marquette Co., *Nichols*.
Lower Peninsula: Alpena Co., *Hermann*. Cheboygan Co., *Nichols*. Leelanau Co., *Darlington*. Livingston and Washtenaw counties, *Steere*.

3. *Tortula ruralis* (Hedw.) Crome — Robust plants in loose or dense tufts, green above, reddish below; leaves squarrose-recurved when moist, oblong-spatulate, rounded-obtuse to truncate or emarginate, ending in a rough hair-point; upper cells densely papillose; spores 10–14 μ, maturing in spring. — On soil or rocks, particularly in calcareous habitats. Greenland; widespread in Canada and much of the United States; Europe, Asia, and North Africa.

ILLUSTRATIONS: Figure 62. Conard, *How to Know the Mosses* (ed. 2), fig. 83d–e. Grout, *Mosses with Hand-Lens and Microscope*, Pl. 35. Welch, *Mosses of Indiana*, fig. 83.

Upper Peninsula: Keweenaw Co., *Gilly & Parmelee*. Marquette Co., *Nichols*. Ontonagon Co., *Nichols & Steere*.
Lower Peninsula: Cheboygan Co., *Nichols*. Emmet Co., *Gilly & Parmelee*.

GRIMMIACEAE

Dark plants in low tufts or extensive mats, mostly on rock; stems erect or prostrate, often freely branched; leaves oblong-lanceolate or oblong-ovate, mostly acute or acuminate, often ending in a hyaline tip or hair-point; costa single, strong; cells ± opaque, mostly subquadrate above; capsules immersed to exserted, mostly erect and symmetric; peristome single, the 16 teeth usually perforate or cleft; calyptra cucullate or mitrate.

Basal cells short, quadrate or oblong, not or only
   slightly sinuose . . . . 1. *Grimmia*
Basal cells linear, with thickened, strongly
   sinuose walls . . . . 2. *Rhacomitrium*

1. *Grimmia* Hedw.

Low, dark-green, grayish, or brownish plants forming cushions or mats; leaves ovate or lanceolate, concave or keeled, acute or acuminate, often awned; margins usually recurved below; costa strong, percurrent or excurrent; upper cells rounded-quadrate; lower cells ± elongate, with smooth or moderately sinuose walls; capsules immersed to exserted, usually ovoid or oblong; peristome of 16 teeth, variously split or perforate.

1. Upper leaves (excluding the perichaetial) not hair-pointed,
   (rarely the tips of some leaves hyaline) 2
1. Upper leaves hair-pointed 8
  2. Apex of some or all leaves obtuse or rounded 3
  2. Apex of upper leaves acute or acuminate 5
3. Leaves channeled above; margins incurved . . .2. *G. unicolor*
3. Leaves flat or keeled; margins plane or revolute 4
    4. Some branches with oblong, obtuse leaves, others
     with linear-lanceolate, acuminate leaves; gemmæ
     usually present at leaf tips; capsules
     ellipsoidal . . . . . . 5. *G. hartmanii* var. *anomala*
    4. Branches uniform; gemmæ none; capsules hemispheric or
     wide-mouthed . . . . . 1b. *G. apocarpa* var. *alpicola*
5. Leaves less than 1.5 mm. long 6
5. Leaves 1.5–3 mm. long 7
    6. Leaves linear-lanceolate; cell walls thick and
     brown; perichaetial leaves
     piliferous . . . . . . 1e. *G. apocarpa* var. *nigrescens*
    6. Leaves lanceolate or ovate-lanceolate; cell walls
     hyaline; perichaetial leaves
     muticous . . . . 1f. *G. apocarpa* var. *dupretii*
7. Most leaves acuminate; capsules oblong; peristome
   orange . . . . 1a. *G. apocarpa* and var. *stricta*
7. Most leaves acute or obtuse; capsules ovate;
   peristome red . . . . . . 1b. *G. apocarpa* var. *alpicola*
    8. Margins inrolled in some part of the leaf, or the
     apex channeled (margins never revolute) 9
    8. Margins plane or erect, or 1 or both recurved 10
9. Leaves drawn out to a long, channeled acumen from an
   ovate base; setae curved . . . . 3. *G. olneyi*
9. Leaves lanceolate or linear-lanceolate;
   setae straight . . . . 4. *G. ovalis*

[73]

10. Most mature leaves 2 mm. long or more **11**
10. Most mature leaves 1–1.5 mm. long **12**
11. Both leaf margins equally and distinctly
    revolute . . . . 1a. *G. apocarpa* and var. *stricta*
11. Both leaf margins plane or only
    1 revolute . . . . 5. *G. hartmanii* var. *anomala*
12. Hair-point 1/3-1/2 the leaf length; leaves linear-
    lanceolate; perichaetial leaves twice as long as other
    leaves . . . . 1c. *G. apocarpa* var. *ambigua*
12. Hair-point shorter; leaves usually lanceolate;
    perichaetial leaves shorter **13**
13. Plants dark-green or black; capsules oblong and
    wide-mouthed . . . . 1. *G. apocarpa*
13. Plants bright-green; capsules
    globose . . . . 1d. *G. apocarpa* var. *conferta*

1. *Grimmia apocarpa* Hedw. — Plants in dark- or brownish-green tufts; leaves ovate-lanceolate, acuminate and gradually narrowed to a hyaline hair-point; setae about 0.5 mm. long; capsules immersed; peristome teeth reddish, papillose; spores in spring to summer. — Common on shaded or sunny rocks. Widespread in North America; nearly cosmopolitan.

ILLUSTRATIONS: Figure 63. Conard, *How to Know the Mosses* (ed. 2), fig. 136a–f. Grout, *Mosses with Hand-Lens and Microscope*, Pl. 18. Jennings, *Mosses of Western Pennsylvania* (ed. 2), Pl. 18. Welch, *Mosses of Indiana*, fig. 84.

Upper Peninsula: Chippewa Co., *Steere*. Keweenaw Co., *Povah*. Mackinac Co., *Nichols*. Ontonagon Co., *Darlington*.
Lower Peninsula: Alpena Co., *Robinson & Wells*. Cheboygan Co., *Nichols*. Shiawassee Co., *Parmelee*. Washtenaw Co., *Steere*.

1a. *Grimmia apocarpa* var. *stricta* (Turn.) Hook. & Tayl. — Leaves lanceolate, acuminate; costa and lamina distinctly papillose at back. — On rocks. Across Canada, south to Arizona and West Virginia; Europe and Asia.

Upper Peninsula: Keweenaw Co., *Hermann*. Ontonagon Co., *Nichols & Steere*.

1b. *Grimmia apocarpa* var. *alpicola* (Hedw.) Hartm. — Leaves ovate-lanceolate, not hair-pointed, ± dentate near the apex. — On rocks, especially along streams. Greenland; across the continent, south to Pennsylvania and Utah; Europe, Africa, and Asia.

ILLUSTRATIONS: Conard, *How to Know the Mosses* (ed. 2), fig. 136g. Grout, *Moss Flora of North America*, vol. 2, Pl. 2.

Upper Peninsula: Keweenaw Co., *Allen & Stuntz*. Marquette Co., *Nichols*. Ontonagon Co., *Darlington*.
Lower Peninsula: Charlevoix Co., *Parmelee*. Cheboygan Co., *Nichols*. Presque Isle Co., *Hermann*.

1c. *Grimmia apocarpa* var. *ambigua* (Sull.) Jones — Small plants (up to 1 cm. high) in gray-green cushions; leaves lanceolate or ovate-lanceolate, the upper with a hyaline, denticulate point 1/3 or more their length; margins plane; perichaetial leaves about twice as long as stem leaves; capsules ellipsoid-cylindric, 1–1.5 mm. long; spores 6–8 μ. — On dry rocks. Widespread but scattered, from New Jersey to Washington and Arizona.

ILLUSTRATION: Grout, *Mosses with Hand-Lens and Microscope*, fig. 54.

Upper Peninsula: Keweenaw Co. (Isle Royale), *Povah*.

1d. *Grimmia apocarpa* var. *conferta* (Funck) Spreng. — Plants in short, fragile cushions 1–2 cm. high; leaves lanceolate or ovate-lanceolate, with a short hair-point, keeled; margins strongly revolute; capsules ovoid-hemispheric, thin-walled; peristome teeth orange-red. — On rocks. Across the continent and south to North Carolina in the East, California in the West; Europe and Asia.

ILLUSTRATION: Grout, *Mosses with Hand-Lens and Microscope*, fig. 53.

Upper Peninsula: Baraga and Houghton counties, *Parmelee*. Keweenaw Co., *Allen* & *Stuntz*. Ontonagon Co., *Darlington*.

1e. *Grimmia apocarpa* var. *nigrescens* Mol. — Plants blackish, with linear-lanceolate leaves; cell walls thick and brown; perichaetial leaves piliferous. — On rocks. Eastern North America; Europe.

Upper Peninsula: Alger Co., *Beach*.

1f. *Grimmia apocarpa* var. *dupretii* (Thér.) Sayre. — Leaves ovate or lanceolate, muticous; margins entire, revolute; cell walls hyaline; capsules ellipsoidal, 0.8 mm. long; peristome smooth, red, entire. — On rocks. Across southern Canada and the northern United States.

Upper Peninsula: Keweenaw Co. (Rock Harbor), *Hermann*.

2. *Grimmia unicolor* Hook. — Plants in dark-green to blackish tufts up to 4 cm. high; leaves channeled and linear-lanceolate from a broad base, up to about 2 mm. long, blunt and muticous; margins incurved; cells 3–4-stratose above; capsules exserted, 1.3–1.5 mm. long, slightly contracted below the mouth; spores yellowish, 10–15 μ, produced in summer. — On moist, siliceous rocks. New England to British Columbia and California.

ILLUSTRATION: Grout, *Mosses with Hand-Lens and Microscope*, fig. 58.

Upper Peninsula: Keweenaw Co., *Holt*. Marquette Co., *Nichols*.

[75]

3. *Grimmia olneyi* Sull. — Leaves 2-3 mm. long, linear-lanceolate from an ovate base, canaliculate-concave, ending in a spinulose hair-point; margins not recurved; basal cells mostly thin- and smooth-walled; spores in early spring. — On non-calcareous rocks. Nova Scotia to the Great Lakes region and Georgia.

ILLUSTRATIONS: Grout, *Mosses with Hand-Lens and Microscope*, Pl. 19. Welch, *Mosses of Indiana*, fig. 88.

Upper Peninsula.: Keweenaw Co., *Povah.*

4. *Grimmia ovalis* (Hedw.) Lindb. — Plants dark-green or grayish; leaves 2–2.5 mm. long, lanceolate, ending in a nearly smooth hair-point about 1/3 the leaf length, concave, bistratose above; margins erect or somewhat incurved; basal cells rectangular with ± thickened and sinuose walls, the basal marginal cells shorter and paler; spores 8–10 μ, in spring. — On rocks. Canada and the northern United States, south to Arizona; Europe and Asia.

ILLUSTRATION: Grout, *Moss Flora of North America*, vol. 2, Pl. 7 (as *G. commutata*).

Upper Peninsula: Keweenaw Co., *Allen & Stuntz.* Ontonagon Co., *Nichols & Steere.*

5. *Grimmia hartmanii Schimp.* var. *anomala* (Hampe) Mönk. — Leaves 1.7–2.5 mm. long, lanceolate from a broader base, muticous or with very short hyaline points; clusters of yellowish propagula common at tips of upper leaves; margins recurved; cells ± papillose; setae curved, 3–5 mm. long; operculum long-rostrate; calyptra mitrate. — On rocks. Alaska to Idaho and Wyoming; Iceland; Europe.

ILLUSTRATION: Grout, *Moss Flora of North America*, vol. 2, Pl. 16.

Upper Peninsula: Keweenaw Co., *Steere.* Marquette Co., *Nichols.*

2. Rhacomitrium Brid.

Plants rather robust, in loose tufts or mats on rocks; stems mostly prostrate, often freely branched; leaves lance-acuminate, mostly ending in a hyaline hair-point, but sometimes blunt or rounded at apex; cells with thick, strongly sinuose walls, narrow and elongate throughout or sometimes shortly-rectangular above; setae straight; capsules erect, ovoid to cylindric; peristome teeth deeply divided into 2–3 filiform segments; operculum conic-rostrate; calyptra mitrate.

1. Leaves without a hyaline tip 2
1. Leaves typically hyaline-tipped 4

[76]

2. Short, tuft-like branches numerous; upper cells
longer than broad . . . . 3. *R. fasciculare*
2. Short branches few or none; upper cells isodiametric . . . 3
3. Upper half of leaves narrowly acuminate; setae mostly
less than 5 mm. long . . . . 1. *R. patens*
3. Upper half of leaves broad, broadly obtuse and dentate
at the apex; setae more than 5 mm. long . . . 2. *R. aciculare*
4. Upper leaf cells elongate 5
4. Upper leaf cells mainly isodiametric 6
5. Leaves smooth; hyaline points sparsely toothed . 6. *R. microcarpon*
5. Leaves papillose; hyaline points strongly toothed . 7. *R. canescens*
6. Short, tuft-like branches numerous; leaves ending
in smooth hyaline points; setae
5–8 mm. long . . . . 4. *R. heterostichum*
6. Short, tuft-like branches none; hyaline points
dentate; setae 2–5 mm. long . . . . . 5. *R. sudeticum*

1. *Rhacomitrium patens* (Hedw.) Hüb. — Leaves lanceolate from an ovate base, muticous and entire or slightly toothed at the apex, 2–3 mm. long, keeled above; costa 2–4-ridged at back; upper cells rounded-quadrate; setae curved, short; capsules ovoid, 1.5–2 mm. long; spores in spring. — On (usually) non-calcareous boulders. Greenland; Alaska to Oregon and Montana; Great Lakes region; Europe and Asia.

ILLUSTRATION: Grout, *Moss Flora of North America*, vol. 2, Pl. 24A.

Upper Peninsula: Keweenaw Co., *Steere*. Marquette Co., *Nichols*.

2. *Rhacomitrium aciculare* (Hedw.) Brid. — Plants dark-green, leafless below; leaves lingulate, broadly rounded and dentate at apex; costa ending below the apex; upper cells short; setae 5–15 mm. long; capsules 2–2.5 mm. long. — On moist or inundated rocks. Widespread in eastern North America; Alaska to California; Europe.

ILLUSTRATIONS: Figure 64. Conard, *How to Know the Mosses* (ed. 2), fig. 133b. Grout, *Mosses with Hand-Lens and Microscope*, Pl. 21. Jennings, *Mosses of Western Pennsylvania* (ed. 2), Pl. 18.

Upper Peninsula: Alger and Keweenaw counties, *Steere*. Marquette Co., *Nichols*. Ontonagon Co., *Steere*.

3. *Rhacomitrium fasciculare* (Hedw.) Brid. — Plants yellow-green to brownish or blackish, with numerous short branches; leaves 2–3 mm. long, linear-lanceolate from an ovate base, narrowly pointed, muticous; margins revolute; upper cells elongate, finely papillose; capsules 2–2.5 mm. long; spores 12–16 μ, in spring. — On moist or wet rocks. Across Canada and Alaska and south to the northern United States.

ILLUSTRATION: Grout, *Mosses with Hand-Lens and Microscope*, Pl. 22.

Upper Peninsula: Marquette Co. (Sugar Loaf Mt.), *Nichols*.

4. Rhacomitrium heterostichum (Hedw.) Brid.

— Plants gray-green, with tuft-like branchlets conspicuous near ends of stems; leaves lanceolate, acuminate, hair-pointed, keeled; margins revolute, unistratose; upper cells quadrate, smooth or nearly so, often bistratose; setae 5–8 mm. long; capsules cylindric to ellipsoidal, 1–3 mm. long. — On rocks. Greenland to Alaska, south to Labrador and California; Europe, Asia, Macaronesia, and Australasia.

ILLUSTRATIONS: Conard, *How to Know the Mosses* (ed. 2), fig. 133c–e. Grout, *Moss Flora of North America*, vol. 2, Pl. 22.

Upper Peninsula: Marquette Co., *Nichols*.
Lower Peninsula: Emmet Co., *Katz*.

5. Rhacomitrium sudeticum (Funck) BSG

— Plants yellow-green or blackish, fasciculately branched; hair-point short or lacking; upper cells short, 2–3-stratose; setae 2–3 mm. long; capsules ovoid, 1–2 mm. long. — On rocks. Greenland to Alaska, south to Pennsylvania and Oregon; Europe, Caucasus, and Japan.

ILLUSTRATION: Grout, *Mosses with Hand-Lens and Microscope*, fig. 60.

Upper Peninsula: Alger Co., *Steere*. Marquette Co., *Nichols*.
Lower Peninsula: Emmet Co., *Steere*.

6. Rhacomitrium microcarpon (Hedw.) Brid.

— Plants yellow-green, ± hoary; leaves oblong-lanceolate, ending in a denticulate, hyaline hair-point; margins revolute, unistratose; cells elongate throughout; setae 4–5 mm. long; capsules 1.8–2 mm. long; spores 10–15 μ, finely granular. — On rocks. Greenland; across the continent, south to Oregon and Georgia; Europe.

ILLUSTRATION: Grout, *Mosses with Hand-Lens and Microscope*, fig. 59.

Upper Peninsula: Chippewa Co., *Thorpe*. Keweenaw Co., *Steere*, Marquette Co., *Nichols*.

7. Rhacomitrium canescens (Hedw.) Brid.

— Plants in yellowish- or grayish-green tufts; leaves lanceolate, with rather broad, hyaline tips or hair-points; upper cells 1–2:1, papillose; setae 5–25 mm. long; capsules ellipsoidal, 1.8–2 mm. long. — On rocks. Greenland; across the continent, south to New Hampshire and California; Europe, Macaronesia, and Asia.

[78]

ILLUSTRATIONS: Conard, *How to Know the Mosses* (ed. 2), fig. 62. Grout, *Mosses with Hand-Lens and Microscope*, fig. 62.

Upper Peninsula: Chippewa Co., *Steere*. Keweenaw Co., *Cooper*. Marquette Co., *Nichols*. Ontonagon Co., *Steere*.

EPHEMERACEAE

Plants minute and inconspicuous, ephemeral, consisting of bud-like clusters of leaves arising from a much branched protonema; leaves lanceolate, slenderly pointed; costa present or absent; cells hexagonal or rhomboidal, lax, sometimes papillose; capsules immersed, subglobose, indehiscent; peristome and operculum none; calyptra minute, mitrate or cucullate.

Ephemerum Hampe

Plants only 1–2 mm. high, gregarious, growing from an abundant protonema; leaves lanceolate, irregularly spinulose; capsules subglobose, apiculate.

Leaf cells and costa ± papillose; calyptrae often papillose; costa
   often excurrent into a long, spinose point .   .   .   1. *E. spinulosum*
Leaves and calyptrae smooth; costa percurrent . . 2. *E. cohaerens*

1. *Ephemerum spinulosum* Bruch & Schimp. — Leaves up to 1.5 mm. long, linear-lanceolate, spinose-serrate, the teeth often recurved; costa often excurrent; cells and back of costa papillose; spores in autumn. — On bare, moist soil. Quebec to Minnesota, south to the Gulf of Mexico.

ILLUSTRATIONS: Conard, *How to Know the Mosses* (ed. 2), fig. 40a–b. Grout, *Moss Flora of North America*, vol. 2, Pl. 27.

Upper Peninsula: Ontonagon Co., *Conard*.
Lower Peninsula: Gratiot Co., *Schnooberger*. Kalamazoo Co., *Becker*. Washtenaw Co., *Steere*.

2. *Ephemerum cohaerens* (Hedw.) Hampe. — Leaves broadly lanceolate to oblong-lanceolate, serrate above; costa ending below the apex to percurrent, often toothed at back; cells smooth; spores 50–70 μ, mature in late autumn to winter. — On moist clay or sand. Eastern North America; Europe.

ILLUSTRATIONS: Figure 65. Conard, *How to Know the Mosses* (ed. 2), fig. 40d.

Lower Peninsula: Washtenaw Co., *Smith*.

Plants in loose tufts or gregarious, with short stems. Leaves usually crowded in apical rosettes, soft, ± concave, ovate or oblong, mostly acute, ± toothed at margins; costa well developed, percurrent or short-excurrent; cells lax and thin-walled, smooth, rectangular, hexagonal, or broadly rhomboidal; capsules mostly exserted, erect to strongly inclined, sometimes asymmetric; peristome, if present, mostly double, the segments opposite the teeth; operculum usually developed; calyptra usually long-beaked, often inflated at base.

1. Capsules inclined and asymmetric; peristome present . 3. *Funaria*
1. Capsules erect and symmetric; peristome none 2
   2. Setae usually long; capsules urn-shaped; exothecial
     cells not collenchymatous . . . . . 2. *Physcomitrium*
   2. Setae virtually lacking; capsules spherical; exothecial
     cells collenchymatous . . . . . . 1. *Aphanorhegma*

1. *Aphanorhegma* Sull.

Very small, gregarious plants with immersed, globose, apiculate capsules; operculum apiculate; exothecial cells collenchymatous; peristome none.

Aphanorhegma serratum (Hook. & Wils.) Sull. — Plants 2–4 mm. high; leaves oblong-lanceolate or oblong-ovate, acute or acuminate, serrulate above; costa ending in or just below the apex; capsules light-brown, often hidden by the leaves; spores 22–30 μ, mature in autumn. — On damp, clayey soil, in shaded or open places. Widespread in eastern North America.

ILLUSTRATIONS: Figure 66. Conard, *How to Know the Mosses* (ed. 2), fig. 106a–d. Grout, *Moss Flora of North America*, vol. 2, Pl. 29. Jennings, *Mosses of Western Pennsylvania* (ed. 2), Pl. 20.

Lower Peninsula: Ingham Co., *Darlington*. Kalamazoo Co., *Becker*. Washtenaw Co., *Steere*.

2. *Physcomitrium* (Brid.) Fürnr.

Rather small plants in loose tufts; leaves crowded, lanceolate, acute or acuminate, serrate above; autoicous; capsules mostly exserted, urn-shaped, erect and symmetric; operculum present; peristome none; calyptra inflated.

Physcomitrium pyriforme (Hedw.) DeNot. — Light-green plants 5–10 mm. high; spores maturing in late spring or early summer. — Common and weedy, on bare soil in open places. Throughout eastern North America; Europe, Macaronesia, North Africa, and the Caucasus.

Pottia truncata, which has similar capsules, is smaller, less common, and fruits in late autumn to spring; also, its leaves are awned.

ILLUSTRATIONS: Figure 67. Conard. *How to Know the Mosses* (ed. 2), fig. 162a–d. Grout, *Mosses with Hand-Lens and Microscope,* fig. 96. Jennings, *Mosses of Western Pennsylvania* (ed. 2), Pl. 20. Welch, *Mosses of Indiana,* fig. 95.

Lower Peninsula: Cheboygan Co., *Roberts.* Clinton Co., *Parmelee.* Ingham Co., *Wheeler.* Leelanau Co., *Darlington.* Washtenaw Co., *Kauffman.*

3. *Funaria* Hedw.

Plants moderately robust, in loose tufts; leaves broadly oblong-ovate or oblong-obovate, acute or acuminate, entire or serrate above; costa percurrent or short-excurrent; autoicous; setae long; capsules inclined and strongly asymmetric, sometimes deeply plicate; peristome double, the segments opposite the teeth; annulus present or absent; operculum present; calyptra inflated.

Funaria hygrometrica Hedw. — Plants light-green or yellowish, 3–10 mm. tall; upper leaves larger and crowded in bud-like clusters; leaves acute or acuminate, entire or slightly serrulate above; setae very long and slender, hygroscopic; capsules furrowed when dry; annulus large and revoluble; spores in late spring or early summer. — A common, weedy moss, growing on soil or other substrata in disturbed areas, particularly in burned-over areas, often in greenhouses. Nearly cosmopolitan.

ILLUSTRATIONS: Figure 68. Conard, *How to Know the Mosses* (ed. 2), fig. 163a–d. Grout, *Mosses with Hand-Lens and Microscope,* fig. 97. Jennings, *Mosses of Western Pennsylvania* (ed. 2), Pl. 21. Welch, *Mosses of Indiana,* fig. 97.

Upper Peninsula: Baraga Co., *Parmelee.* Chippewa Co., *Thorpe.* Keweenaw Co., *Cooper.* Ontonagon Co., *Darlington.*

Lower Peninsula: Alpena Co., *Wheeler.* Cheboygan Co., *Nichols.* Eaton Co., *Darlington.* Ingham Co., *Parmelee.* Leelanau Co., *Darlington.* Lenawee Co., *Stearns.* Washtenaw Co., *Kauffman.*

SPLACHNACEAE

Plants mostly growing in tufts on dung or animal remains, erect; leaves soft, lanceolate or obovate; costa mostly ending near the apex; cells large and lax, smooth, oblong-hexagonal above, rectangular below; setae elongate; capsules erect and symmetric, with a well-developed, sometimes conspicuously inflated hypophysis; operculum present; peristome teeth generally present, 16, often joined in 2's or 4's.

Hypophysis not or only slightly wider than the urn . 1. *Tetraplodon*
Hypophysis much wider than the urn . . . . . 2. *Splachnum*

1. *Tetraplodon* BSG

Leaves lanceolate, acuminate, entire or serrate; costa ending in or below the apex; cells mostly rectangular, thin-walled or sometimes thick-walled above; hypophysis narrow but elongate and conspicuous; peristome teeth ± in 4's, later in 2's, reflexed.

Tetraplodon angustatus (Hedw.) BSG — Small plants (5–10 mm. tall); leaves narrow, long-subulate-acuminate, irregularly serrate above; costa ending in the subula (and appearing to be excurrent); urn brown, the hypophysis paler, about twice as long; spores 7–11 μ, maturing in late summer. — On dung or decaying carcasses of small mammals. Greenland; across the continent and south to British Columbia and New York; Europe and Asia.

ILLUSTRATIONS: Figure 69. Grout, *Moss Flora of North America*, vol. 2, Pl. 41. Jennings, *Mosses of Western Pennsylvania* (ed. 2), Pl. 72.
Upper Peninsula: "Lake Superior region," *Cooley*.

2. *Splachnum* Hedw.

Leaves broadly ovate or ovate-lanceolate, soft, entire or serrate; costa mostly ending below the apex; cells large and thin-walled, oblong-hexagonal above; setae elongate, slender; hypophysis much larger and much wider than the urn, darker or sometimes brightly colored (red or yellow); peristome teeth united in pairs, sometimes becoming free.

Leaves sparsely toothed at apex; hypophysis
  top-shaped . . . . 1. *S. ampullaceum*
Leaves serrate nearly to the base; hypophysis much inflated,
  forming a purple-red, flaring skirt . . . . . 2. *S. rubrum*

1. *Splachnum ampullaceum* Hedw. — Leaves light-green, obovate to acuminate, irregularly dentate at the apex; hypophysis large, top-shaped, becoming pale-purple; spores in summer. — On dung (particularly of moose). Newfoundland to Wisconsin, south to New Jersey and Ohio; Europe and Asia.

ILLUSTRATIONS: Figure 70. Conard, *How to Know the Mosses* (ed. 2), fig. 52a–d. Grout, *Mosses with Hand-Lens and Microscope*, fig. 94. Jennings, *Mosses of Western Pennsylvania* (ed. 2), Pl. 52.
Upper Peninsula: Keweenaw Co. (Isle Royale), *Cooper*.

2. *Splachnum rubrum* Hedw. — Light-green plants about 2–4 cm. high; leaves ovate-lanceolate, acuminate (the upper leaves tapered to a long, slender point), serrate nearly to the base; hypophysis wide and skirt-like or umbrella-like, dark-red or purplish; spores in June or July. — On dung (especially of moose). Maine and New Brunswick to Saskatchewan and the Rocky Mts., rare; Europe and northern Asia.

ILLUSTRATION: Grout, *Moss Flora of North America*, vol. 2, Pl. 43.

Upper Peninsula: Keweenaw Co. (Isle Royale), *Povah*.

SCHISTOSTEGACEAE

Plants small, delicate, and inconspicuous, growing on soil in dark places from a persistent protonema which reflects light from groups of globose cells at the ends of its branches; sterile plants erect, naked below, frondiform above, with oblong-lanceolate, plane, decurrent and confluent, ecostate leaves in 2 rows; fertile stems with lanceolate leaves in a terminal cluster; cells elongate, lax, smooth; dioicous; setae 2–4 mm. long; capsules minute, erect, oblong-ovoid; peristome and annulus none; operculum flat. A family of a single genus, *Schistostega* Mohr and a single species: *S. pennata* Hook. & Tayl. — Plants up to 10 mm. high; spores smooth, 8–10 μ, maturing in spring. This moss has a golden-green glow, as seen from the mouth of caves or other dark holes (under old stumps and overturned bases of trees, for example). Grout said it looks something like a cat's eyes in the dark. — Rare; New England to British Columbia; Europe and Asia.

ILLUSTRATIONS: Figure 71. Grout, *Mosses with Hand-Lens and Microscope*, fig. 93 (as *S. osmundacea*).

Upper Peninsula: Alger Co. (near Munising), *Nichols*.

BRYACEAE

Plants mostly tufted, small to fairly large; stems erect, simple or forked; leaves generally ovate-lanceolate, often crowded at the apex of the stem, often bordered by linear cells; costa single, well developed; cells mostly oblong-hexagonal, rather large and fairly lax, smooth; setae elongate; capsules ovoid to cylindric or pyriform, often with a well-developed neck, mostly inclined to pendulous; peristome mostly double and consisting of 16 lanceolate teeth and an endostome with segments alternating with the teeth, basal membrane well developed, cilia often present; calyptra cucullate.

[83]

1. Peristome single, consisting of 16 linear
   segments . . . . 1. *Mielichhoferia*
1. Peristome double 2
   2. Leaf cells long and narrow (4:1 or longer) 3
   2. Leaf cells less than 4:1 4
3. Capsules pyriform; leaves setaceous . . . . 2. *Leptobryum*
3. Capsules clavate to short- or long-cylindric;
   leaves broader . . . . 3. *Pohlia*
   4. Plants erect from a horizontal, underground
      "rhizome;" setae clustered . . . . . . 5. *Rhodobryum*
   4. Plants not connected by a rhizome; setae single . . 4. *Bryum*

1. *Mielichhoferia* Hornsch.

Small plants in loose tufts; leaves lanceolate; costa ending in or
below the apex; cells narrow and elongate; sporophytes produced in
basal buds; setae elongate; capsules erect to inclined, clavate; annulus large; outer peristome none; endostome of 16 linear, papillose
segments from a low basal membrane.

Mielichhoferia mielichhoferiana (Funck) Limpr. — Low, yellow-
green plants; stems very short, greatly exceeded by subfloral innovations which make the inflorescences appear lateral and basal;
capsules suberect. — On rocks. Rare but widely scattered; Alaska,
California, Maine, Ontario, Michigan, and Tennessee; Europe and
the Caucasus.

ILLUSTRATIONS: Figure 72. Grout, *Moss Flora of North America*, vol. 2, Pl.
71B.
Upper Peninsula: Alger Co., *Nichols.* Keweenaw Co., *Steere.*

2. *Leptobryum* (BSG) Schimp.

Loosely tufted plants with wide-spreading, setaceous leaves; cells
linear; capsules pyriform, inclined to pendulous; peristome double,
basal membrane of endostome high, segments perforated, cilia appendiculate.

Leptobryum pyriforme (Hedw.) Schimp. — Small, yellow-green
plants; capsules shiny; spores in early summer. — On damp soil or
rocks, often weedy. Nearly cosmopolitan.

ILLUSTRATIONS: Figure 73. Conard, *How to Know the Mosses* (ed. 2), fig. 123.
Grout, *Mosses with Hand-Lens and Microscope*, fig. 105. Jennings, *Mosses of
Western Pennsylvania* (ed. 2), Pl. 21. Welch, *Mosses of Indiana*, fig. 116.

Upper Peninsula: Alger Co., *Wheeler.* Chippewa Co., *Thorpe.* Keweenaw
Co., *Hermann.* Marquette Co., *Nichols.* Ontonagon Co., *Darlington.*

Lower Peninsula: Cheboygan Co., *Nichols.* Emmet Co., *Ehlers.* Ingham Co.,
Parmelee. Leelanau Co., *Darlington.* Van Buren Co., *Kauffman.*

[84]

Illustrations

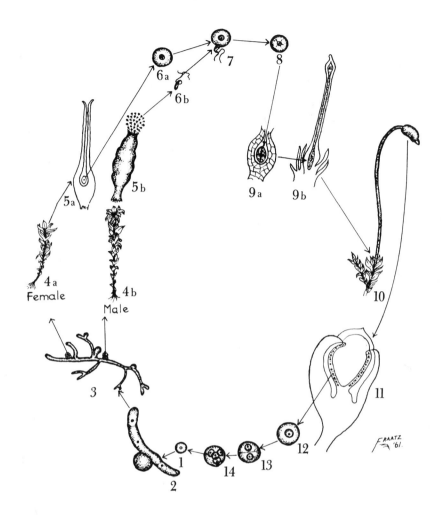

FIGURE 1. The life cycle of a typical moss—1. Spore. 2. Germinating spore. 3. Protonema with buds. 4. Male and female gametophytes. 5a. Archegonium with mature egg. 5b. Antheridium discharging sperms. 6a. Egg. 6b. Sperm. 7. Fertilization. 8. Zygote. 9. Early growth of the young sporophyte inside the archegonium. 10. Mature sporophyte attached to the gametophyte and consisting of a foot, seta, and capsule. 11. Capsule in longitudinal section showing the spore-mother cells. 12-14. Reduction division and formation of four haploid spores from one diploid spore-mother cell. *(Reprinted from Museum Service, Bulletin of Rochester Museum of Arts and Sciences, by permission.)*

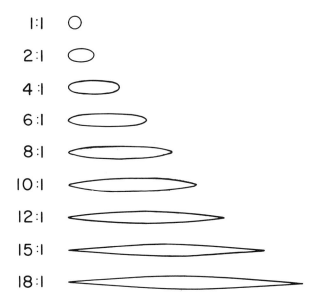

FIGURE 2. Length : breadth ratios often used in descriptions.

FIGURE 3. *Sphagnum magellanicum.* a. Branch leaf in section. b. Branch cortical cells. c. Retort cells. d. Stem cortical cells. FIGURE 4. *Sphagnum centrale.* a. Branch leaf sections. b. Branch leaf. c. Stem leaf. FIGURE 5. *Sphagnum papillosum.* a. Branch leaf section. b. Stem leaf. c. Stem cortical cells. FIGURE 6. *Sphagnum palustre.* a. Branch leaf section. b. Stem cortical cells. FIGURE 7. *Sphagnum compactum.* a. Branch leaf sections. b. Stem cortical cells. c. Retort cells. d. Stem leaf.

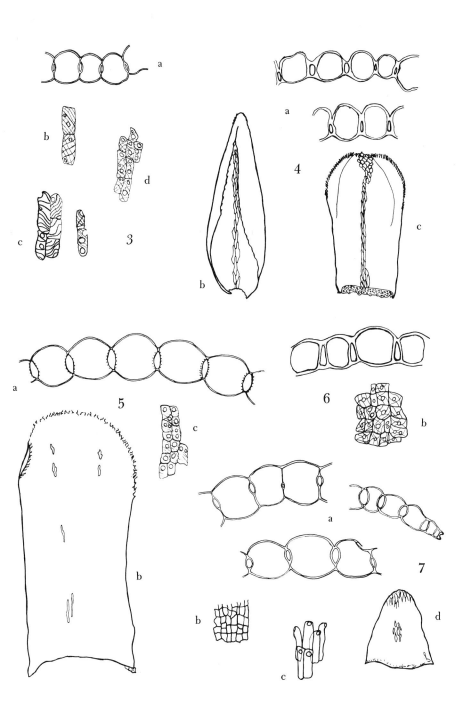

FIGURE 8. *Sphagnum wulfianum*. a. Branch leaf sections. b. Branch leaf. c. Median cells of branch leaf, outer surface. d. Median cells of branch leaf, inner surface. e. Portion of stem showing fascicles of branches. FIGURE 9. *Sphagnum squarrosum*. a. Branch leaf sections. b. Stem leaf. FIGURE 10. *Sphagnum teres*. a. Branch leaf sections. b. Median cells of branch leaf, outer surface (papillae very fine). c. Stem leaf. FIGURE 11. *Sphagnum lindbergii*. a. Branch leaf sections. b. Stem leaf. c. Segment of stem cross-section.

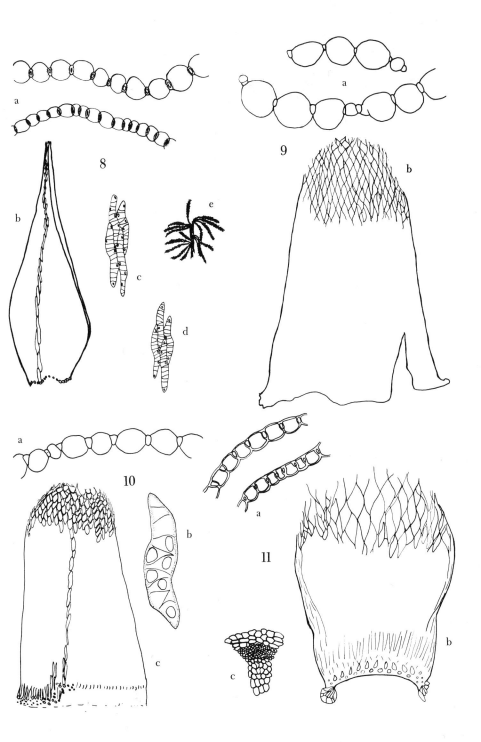

a

8

b

c

d

e

a

9

b

a

10

b

c

a

11

b

c

FIGURE 12A. *Sphagnum recurvum.* a. Branch leaf section. b. Branch leaf. c. Segment of stem cross-section. d. Portion of branch showing undulate leaves. FIGURE 12B. *Sphagnum recurvum* var. *tenue.* a. Branch leaf. b. Portion of branch showing recurved leaf tips (when dry). FIGURE 13. *Sphagnum pulchrum.* a. Branch leaf section. b. Retort cells. c. Apical cells of branch leaf, inner surface. FIGURE 14A. *Sphagnum cuspidatum.* a. Branch leaf sections. b. Branch leaf. c. Stem leaf. FIGURE 14B. *Sphagnum cuspidatum* var. *serrulatum.* Apex of branch leaf. FIGURE 15. *Sphagnum dusenii.* a. Branch leaf sections. b. Branch leaf cells, outer surface. c. Retort cells. d. Stem leaf.

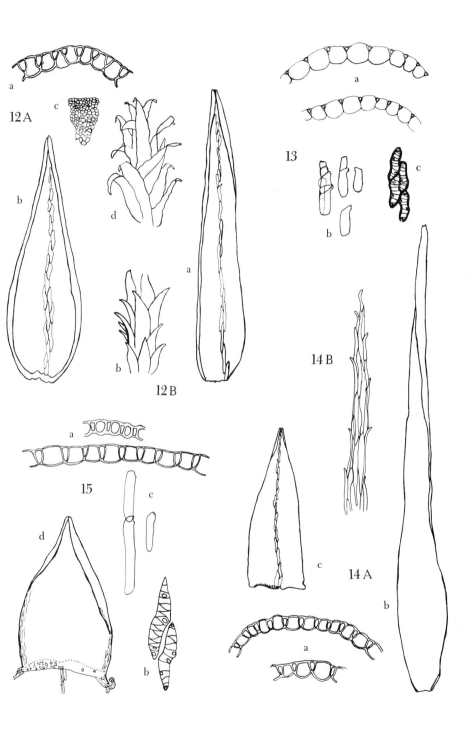

12 A

a

c

d

b

b

12 B

a

13

b

c

14 B

14 A

c

a

b

15

a

c

d

b

b

FIGURE 16. *Sphagnum subsecundum.* a. Branch leaf sections. b. Branch leaf cells, outer surface. c. Stem leaf. FIGURE 17. *Sphagnum fimbriatum.* a. Branch leaf sections. b. Stem leaf. FIGURE 18. *Sphagnum girgensohnii.* a. Branch leaf sections. b. Branch leaf. c. Cells of branch leaf, outer surface. d. Stem leaf. FIGURE 19. *Sphagnum robustum.* a. Branch leaf sections. b. Cells of branch leaf, outer surface. c. Stem leaf.

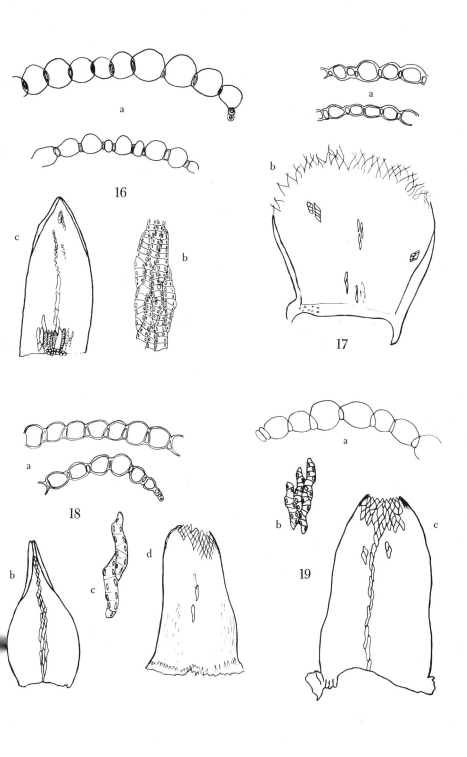

16

17

18

19

FIGURE 20. *Sphagnum fuscum.* a. Branch leaf sections. b. Branch leaf. c. Cells of branch leaf, outer surface. d. Stem leaf. FIGURE 21. *Sphagnum warnstorfianum.* a. Branch leaf section. b. Cells of branch leaf near apex, outer surface. c. Retort cells. d. Segment of stem in cross-section. e. Stem leaf. FIGURE 22A. *Sphagnum capillaceum.* a. Branch leaf sections. b–c. Stem leaves. FIGURE 22B. *Sphagnum capillaceum* var. *tenellum,* stem leaf. FIGURE 23. *Sphagnum plumulosum.* a. Branch leaf sections. b. Upper cells of branch leaf, outer surface. c. Apical cells of branch leaf, inner surface. d. Stem leaf. FIGURE 24. *Sphagnum tenerum.* a. Branch leaf sections. b. Median cells of branch leaf, outer surface. c. Stem leaf. d. Median cells of stem leaf, outer surface.

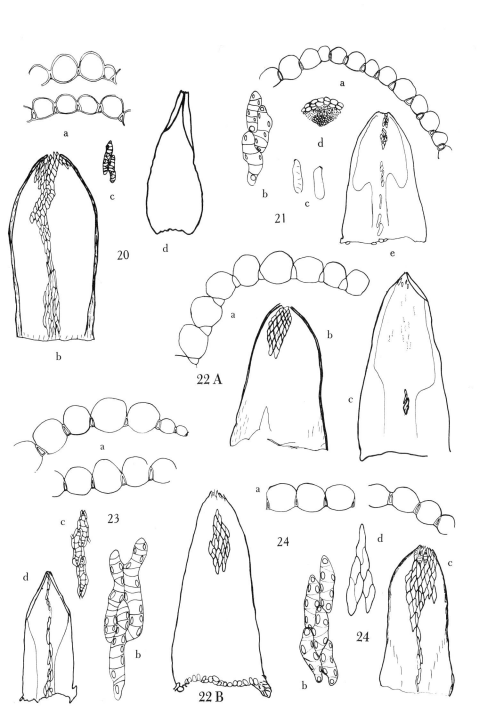

20

a

c

b

d

21

a

b

c

d

e

22 A

a

b

c

23

a

c

b

d

22 B

24

a

b

24

c

d

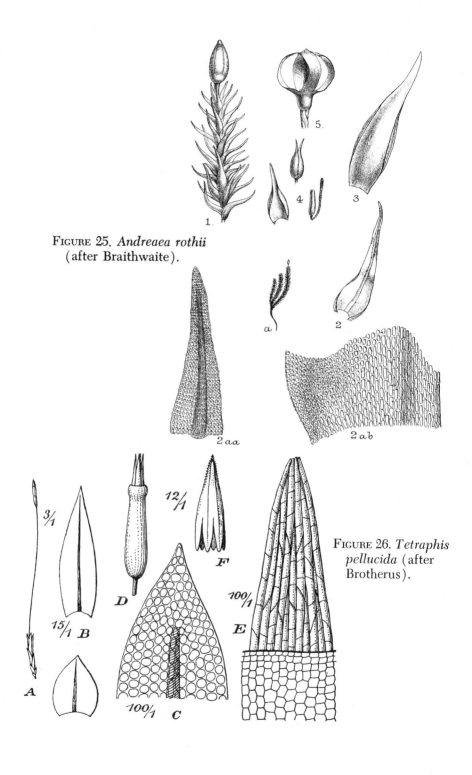

FIGURE 25. *Andreaea rothii* (after Braithwaite).

FIGURE 26. *Tetraphis pellucida* (after Brotherus).

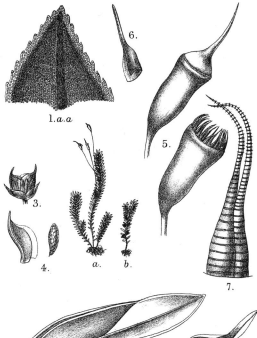

FIGURE 27. *Fissidens*
cristatus
(after Braithwaite).

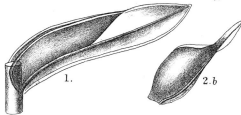

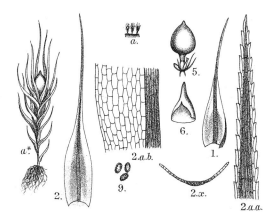

FIGURE 28. *Pleuridium*
subulatum
(after Braithwaite).

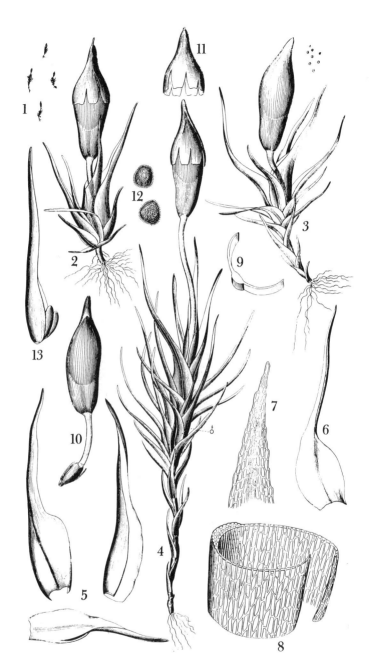

FIGURE 29. *Bruchia sullivantii* (after Sullivant).

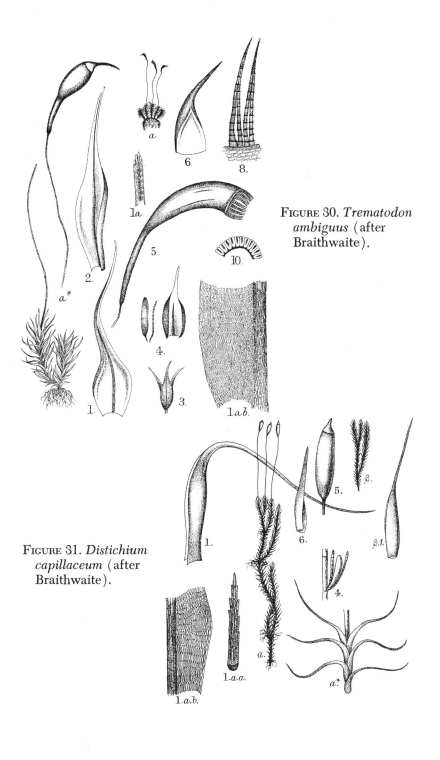

FIGURE 30. *Trematodon ambiguus* (after Braithwaite).

FIGURE 31. *Distichium capillaceum* (after Braithwaite).

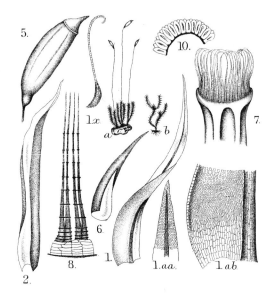

FIGURE 32. *Ceratodon*
purpureus
(after Braithwaite).

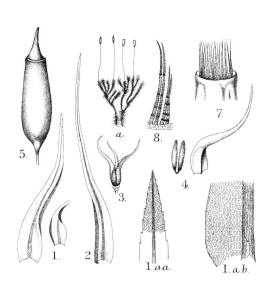

FIGURE 33. *Sælania*
glaucescens
(after Braithwaite).

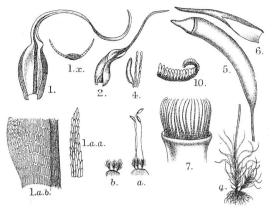

FIGURE 34. *Trichodon cylindricus* (after Braithwaite).

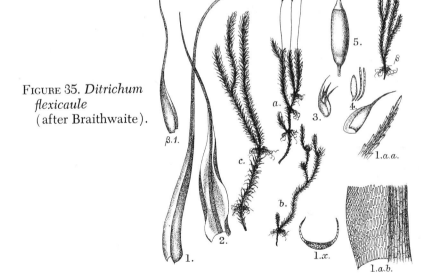

FIGURE 35. *Ditrichum flexicaule* (after Braithwaite).

FIGURE 36. *Seligeria calcarea*
(after Braithwaite).

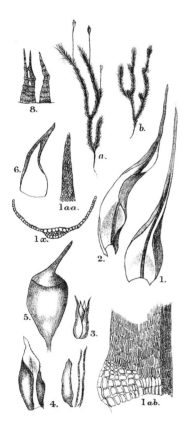

FIGURE 37. *Blindia acuta*
(after Braithwaite).

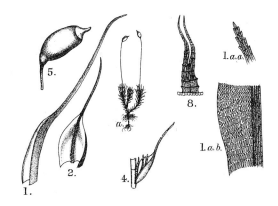

FIGURE 38. *Dicranella heteromalla* (after Braithwaite).

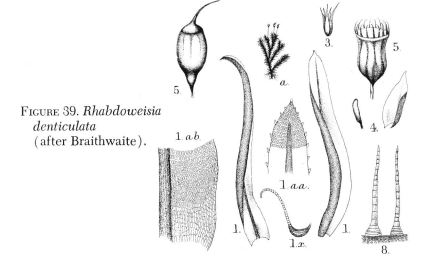

FIGURE 39. *Rhabdoweisia denticulata* (after Braithwaite).

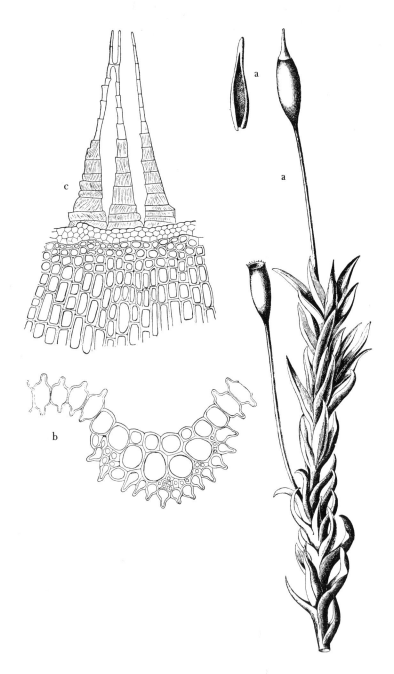

FIGURE 40. *Oreoweisia serrulata* (after Braithwaite).

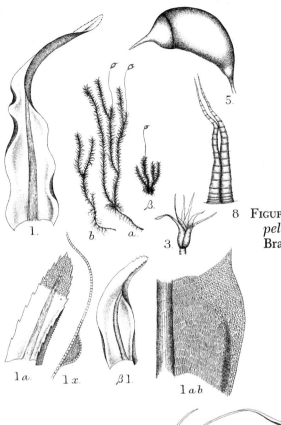

8 FIGURE 41. *Dichodontium pellucidum* (after Braithwaite).

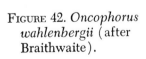

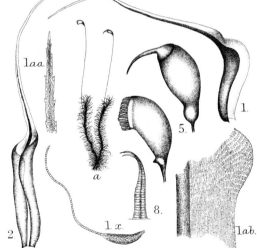

FIGURE 42. *Oncophorus wahlenbergii* (after Braithwaite).

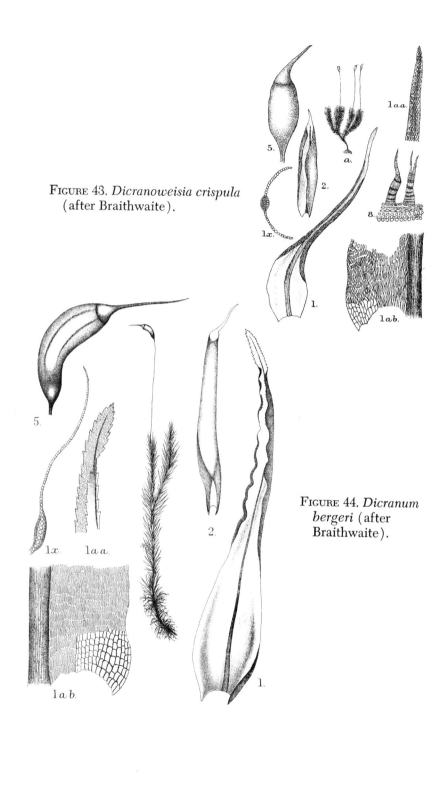

FIGURE 43. *Dicranoweisia crispula* (after Braithwaite).

FIGURE 44. *Dicranum bergeri* (after Braithwaite).

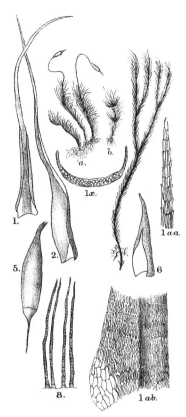

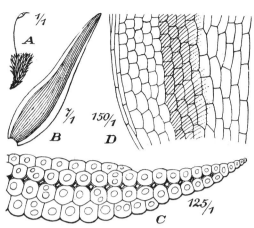

FIGURE 46. *Leucobryum glaucum*
(after Brotherus).

FIGURE 45. *Dicranodontium denudatum*
(after Braithwaite).

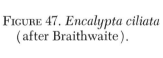

FIGURE 47. *Encalypta ciliata*
(after Braithwaite).

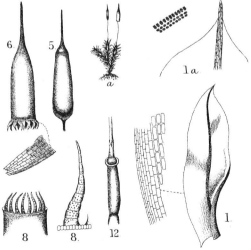

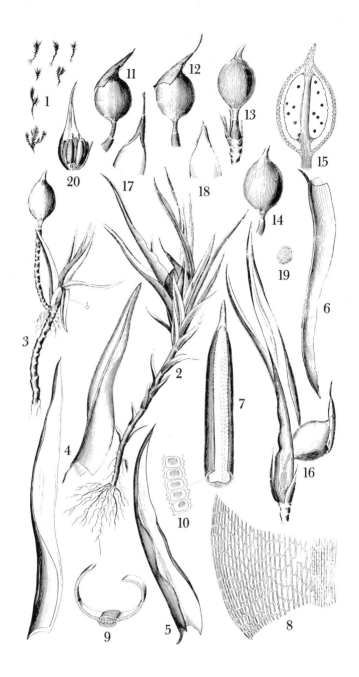

FIGURE 48. *Astomum muehlenbergianum* (after Sullivant).

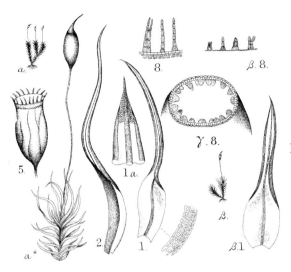

FIGURE 49. *Weissia controversa* (after Braithwaite).

FIGURE 50. *Gymnostomum æruginosum* (after Braithwaite).

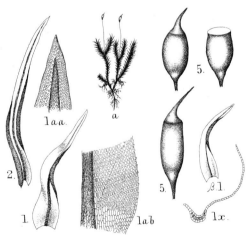

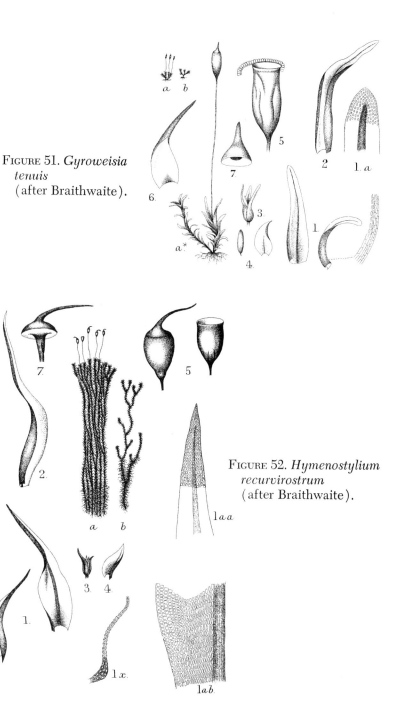

FIGURE 51. *Gyroweisia
tenuis*
(after Braithwaite).

FIGURE 52. *Hymenostylium
recurvirostrum*
(after Braithwaite).

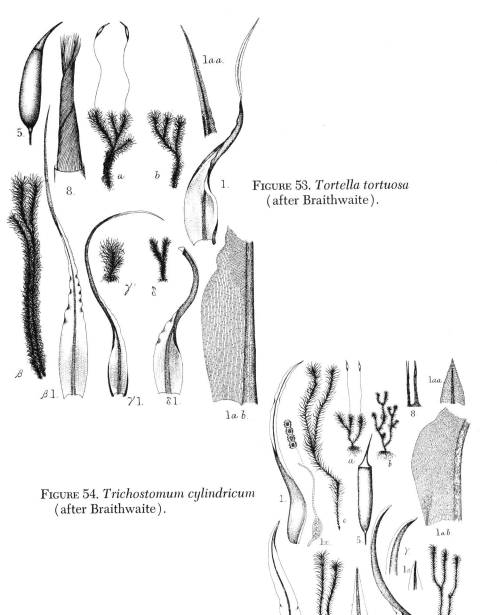

FIGURE 53. *Tortella tortuosa*
(after Braithwaite).

FIGURE 54. *Trichostomum cylindricum*
(after Braithwaite).

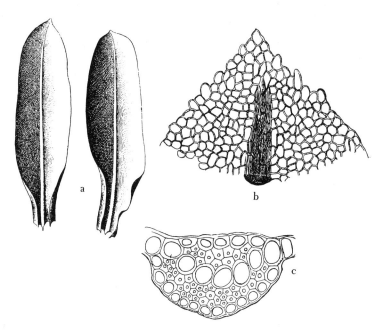

FIGURE 55. *Hyophila tortula* (after Limpricht).

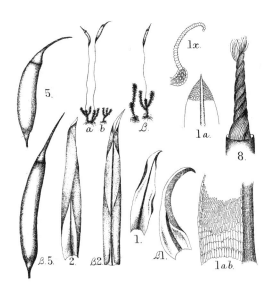

FIGURE 56. *Barbula convoluta* (after Braithwaite).

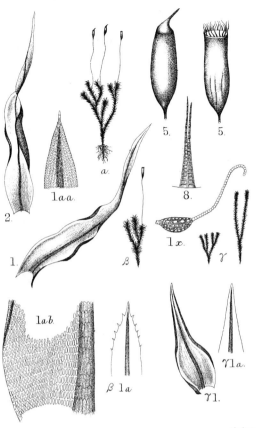

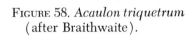

FIGURE 57. *Didymodon
recurvirostris*
(after Braithwaite).

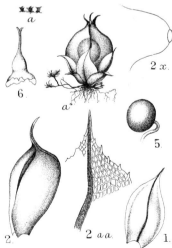

FIGURE 58. *Acaulon triquetrum*
(after Braithwaite).

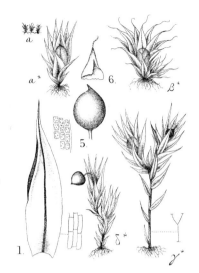

FIGURE 59. *Phascum cuspidatum* (after Braithwaite).

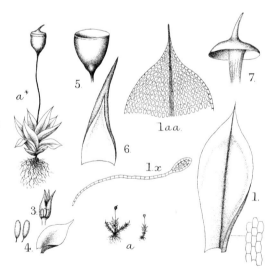

FIGURE 60. *Pottia truncata* (after Braithwaite).

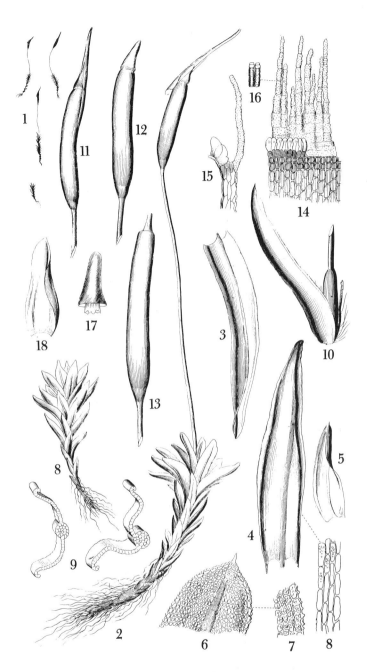

FIGURE 61. *Desmatodon obtusifolius* (aftter Sullivant).

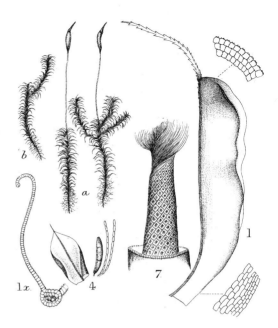

FIGURE 62. *Tortula ruralis*
(after Braithwaite).

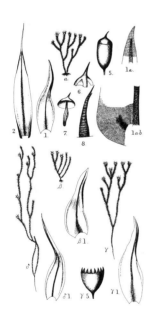

FIGURE 63. *Grimmia apocarpa*
(after Braithwaite).

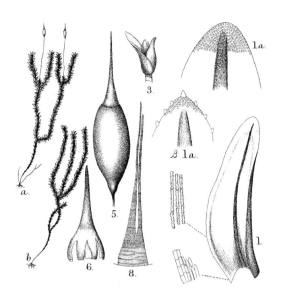

FIGURE 64. *Rhacomitrium aciculare* (after Braithwaite).

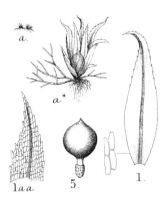

FIGURE 65. *Ephemerum cohaerens* (after Braithwaite).

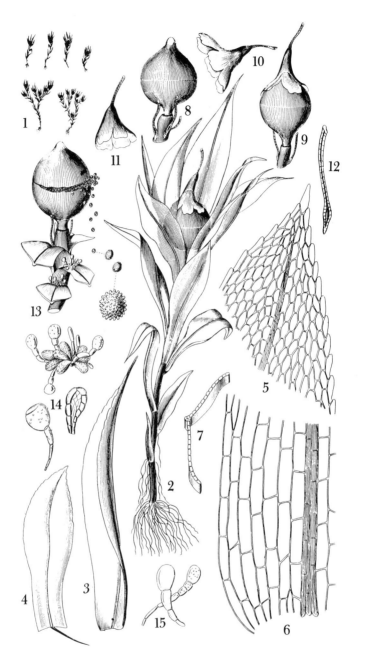

FIGURE 66. *Aphanorhegma serratum* (after Sullivant).

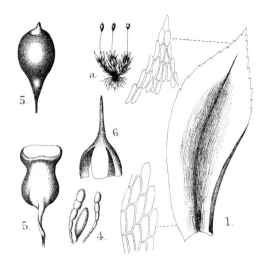

FIGURE 67. *Physcomitrium pyriforme* (after Braithwaite).

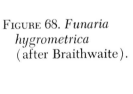

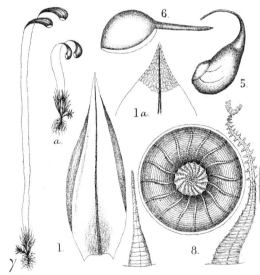

FIGURE 68. *Funaria hygrometrica* (after Braithwaite).

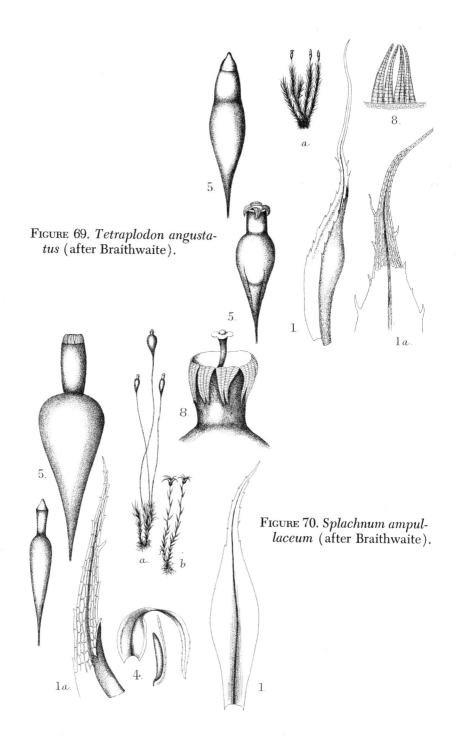

FIGURE 69. *Tetraplodon angusta-tus* (after Braithwaite).

FIGURE 70. *Splachnum ampul-laceum* (after Braithwaite).

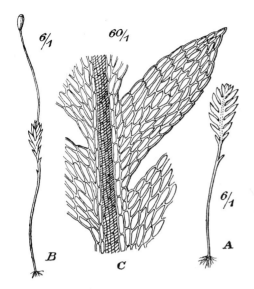

FIGURE 71. *Schistostega
pennata*
(after Brotherus).

FIGURE 72. *Mielichhoferia
mielichhoferiana*
(after Braithwaite).

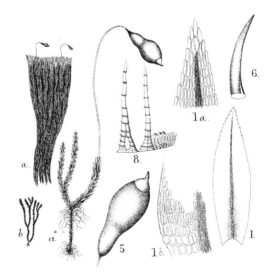

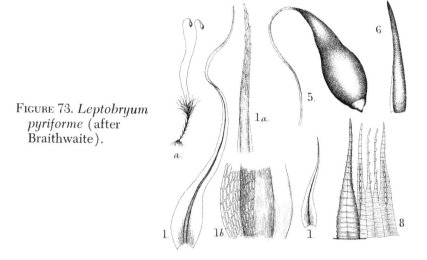

FIGURE 73. *Leptobryum pyriforme* (after Braithwaite).

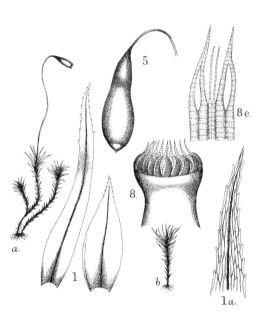

FIGURE 74. *Pohlia cruda* (after Braithwaite).

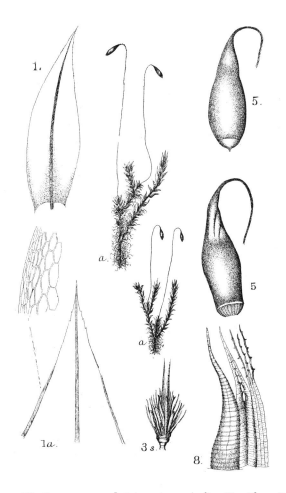

FIGURE 75. *Bryum pseudotriquetrum* (after Braithwaite).

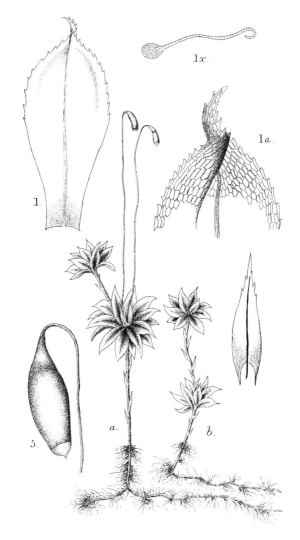

FIGURE 76. *Rhodobryum roseum* (after Braithwaite).

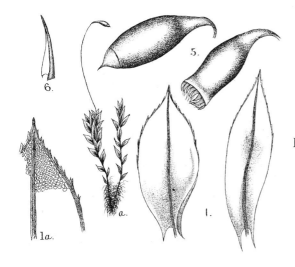

Figure 77. *Mnium serratum* (after Braithwaite).

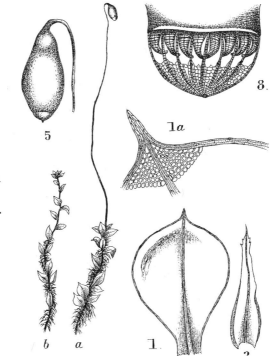

Figure 78. *Cinclidium stygium* (after Braithwaite).

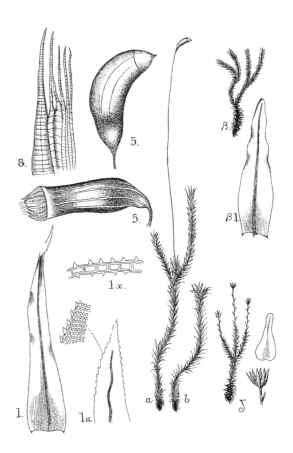

FIGURE 79. *Aulacomnium palustre* (after Braithwaite).

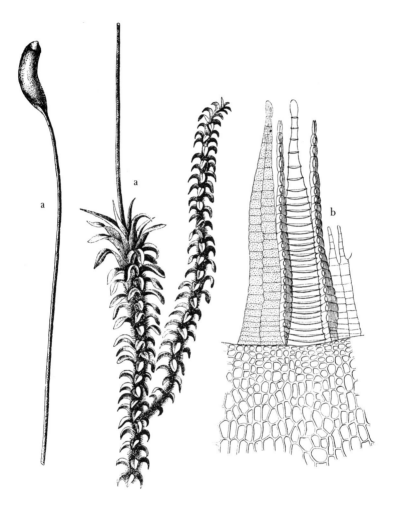

FIGURE 80. *Paludella squarrosa* (after Limpricht).

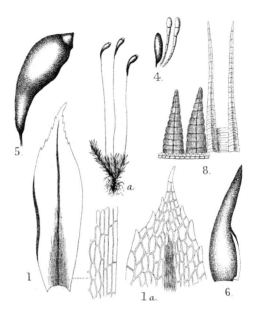

FIGURE 81. *Amblyodon dealbatus* (after Braithwaite).

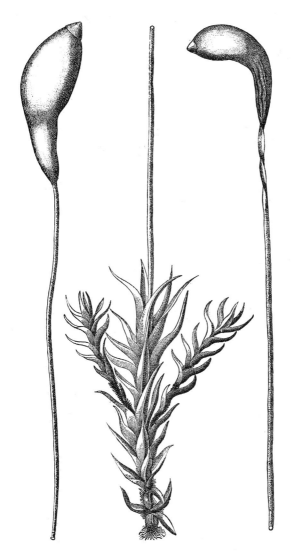

FIGURE 82. *Meesia tristicha* (after Limpricht).

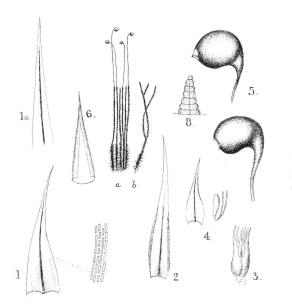

FIGURE 83. *Catoscopium nigritum* (after Braithwaite).

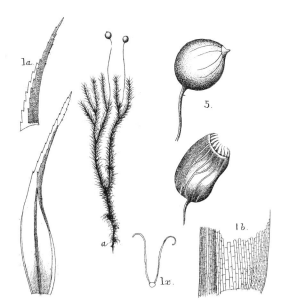

FIGURE 84. *Plagiopus oederi* (after Braithwaite).

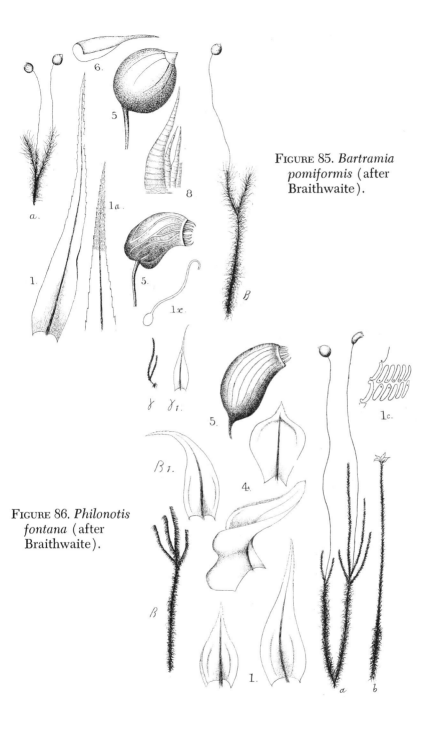

Figure 85. *Bartramia pomiformis* (after Braithwaite).

Figure 86. *Philonotis fontana* (after Braithwaite).

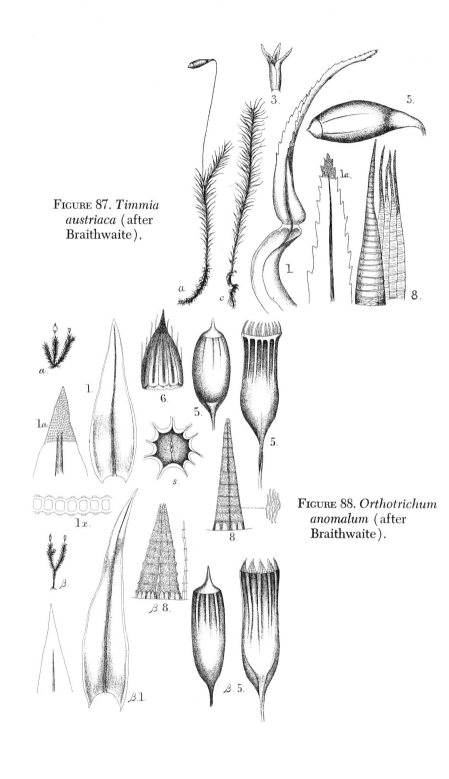

FIGURE 87. *Timmia austriaca* (after Braithwaite).

FIGURE 88. *Orthotrichum anomalum* (after Braithwaite).

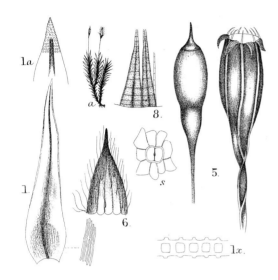

FIGURE 89. *Ulota
hutchinsiæ*
(after Braithwaite)

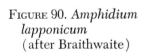

FIGURE 90. *Amphidium
lapponicum*
(after Braithwaite)

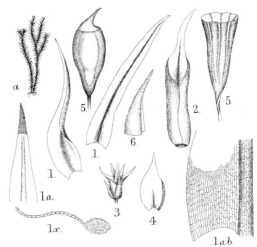

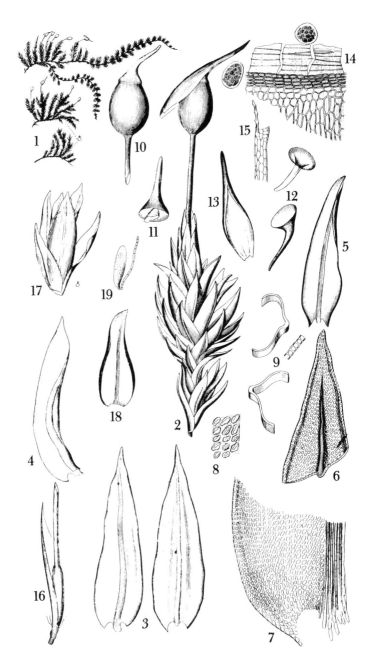

FIGURE 91. *Drummondia prorepens* (after Sullivant).

FIGURE 92. *Fontinalis antipyretica* (after Limpricht).

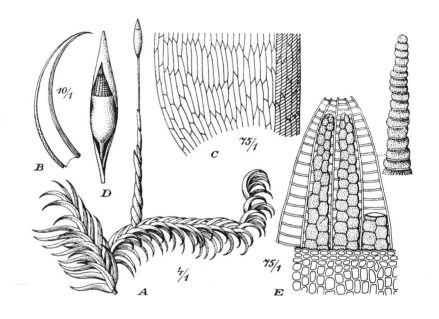

FIGURE 93. *Dichelyma falcatum* (after Brotherus).

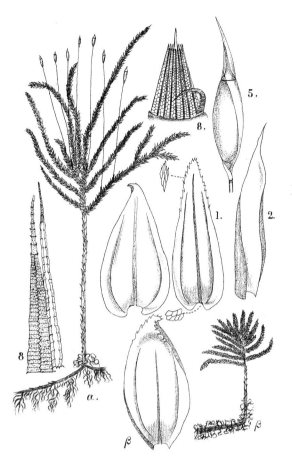

FIGURE 94. *Climacium dendroides* (after Braithwaite)

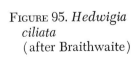

FIGURE 95. *Hedwigia ciliata* (after Braithwaite)

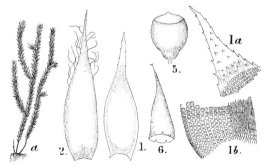

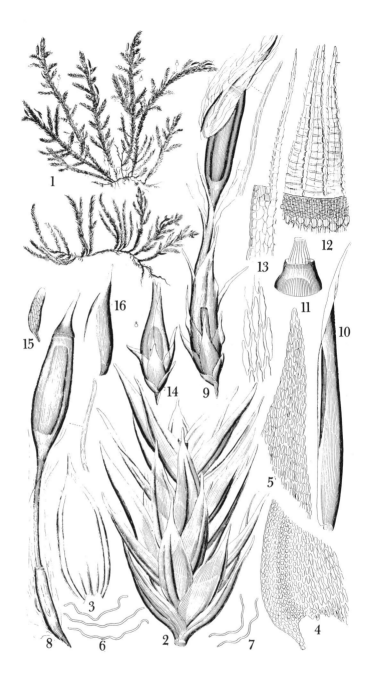

FIGURE 96. *Forsstroemia trichomitria* (after Sullivant).

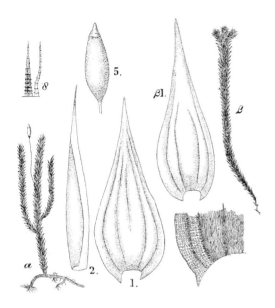

FIGURE 97. *Leucodon*
sciuroides
(after Braithwaite)

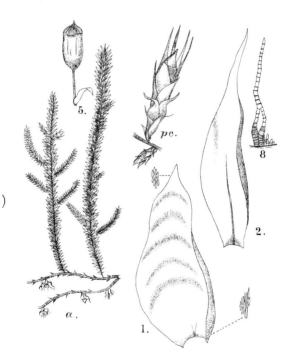

FIGURE 98. *Neckera*
pennata
(after Braithwaite)

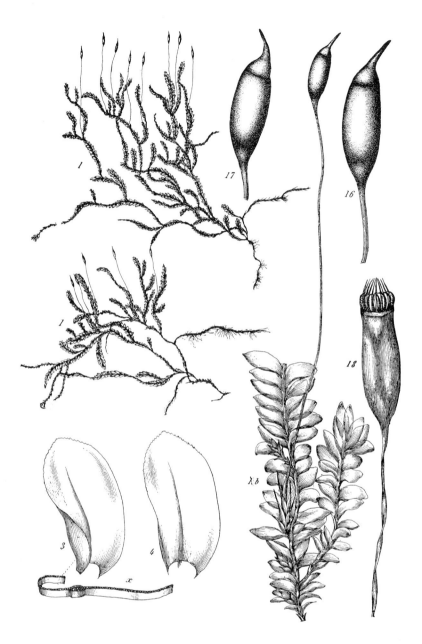

FIGURE 99. *Homalia jamesii* (after *Bryologia Europæa*, as *H. trichomanoides*, which is doubtfully distinct from *H. jamesii*).

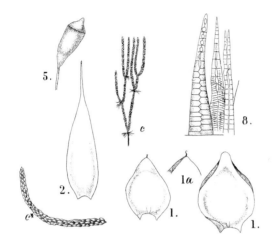

FIGURE 100. *Myurella julacea* (after Braithwaite).

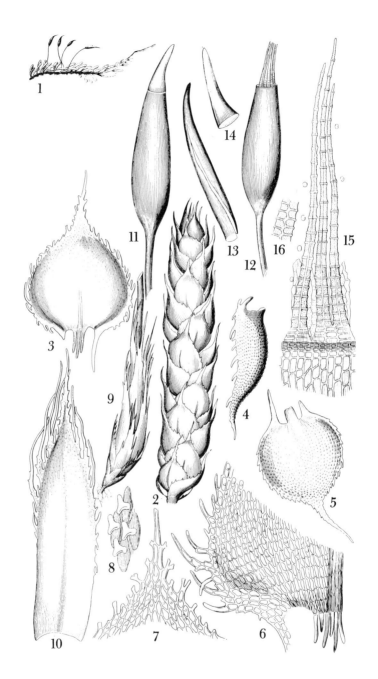

FIGURE 101. *Thelia asprella* (after Sullivant).

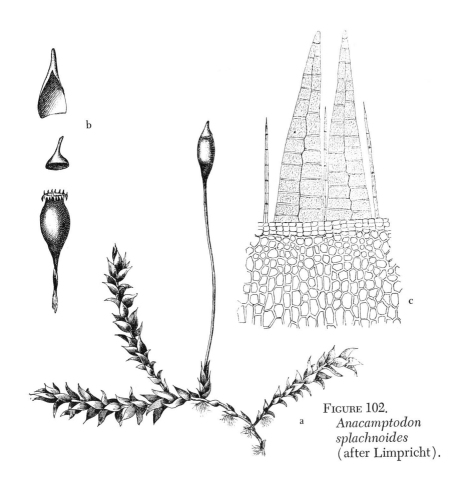

b

c

a

FIGURE 102.
Anacamptodon
splachnoides
(after Limpricht).

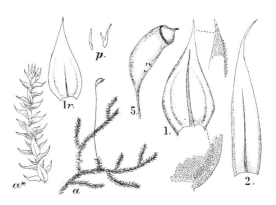

FIGURE 103. *Pseudoleskea*
patens
(after Braithwaite)

*a**

a

p.

1*r.*

5.

1.

2.

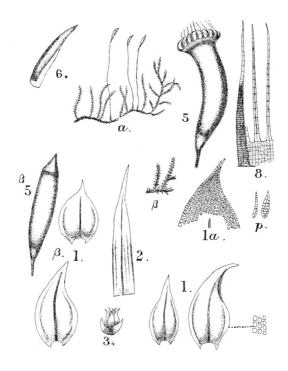

FIGURE 104. *Leskea polycarpa* (after Braithwaite).

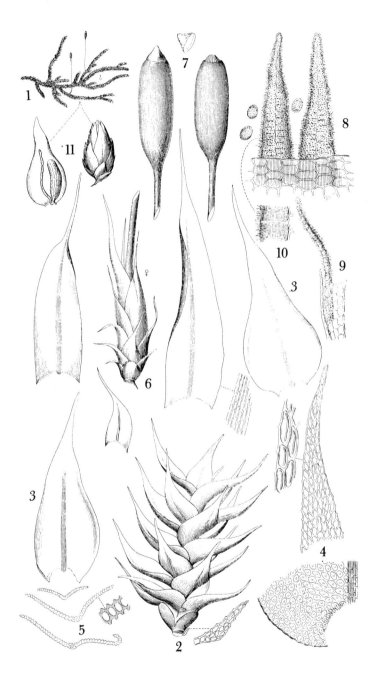

FIGURE 105. *Lindbergia brachyptera* (after Sullivant).

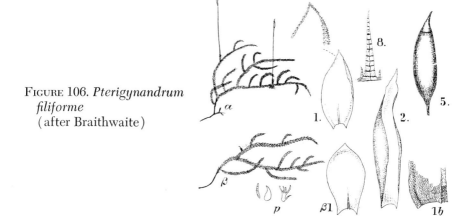

FIGURE 106. *Pterigynandrum*
filiforme
(after Braithwaite)

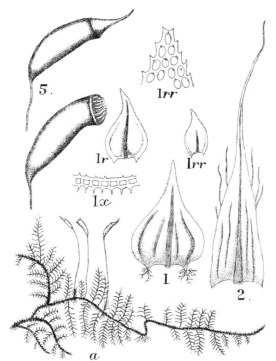

FIGURE 107. *Thuidium*
delicatulum
(after Braithwaite)

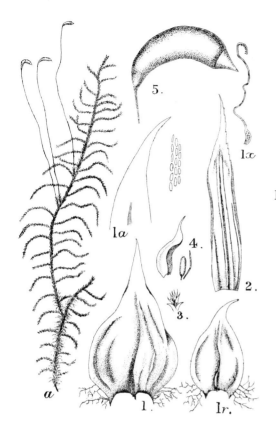

5.

1x

1a

4.

2.

3.

a

1.

1r.

Figure 108. *Helodium blandowii* (after Braithwaite)

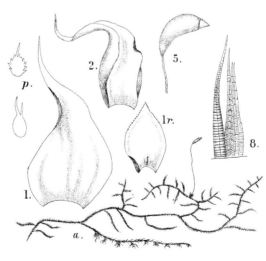

Figure 109. *Heterocladium squarrosulum* (after Braithwaite).

p.

2.

5.

1r.

8.

1.

a.

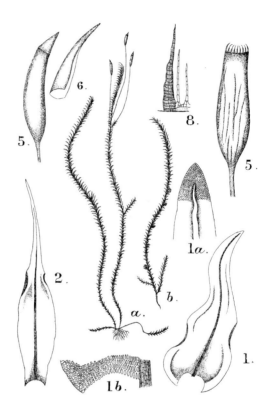

FIGURE 110. *Anomodon viticulosus* (after Braithwaite).

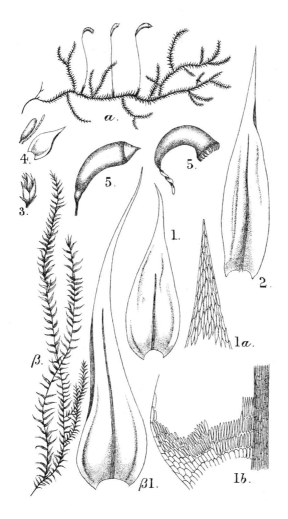

FIGURE 111. *Leptodictyum riparium* (after Braithwaite).

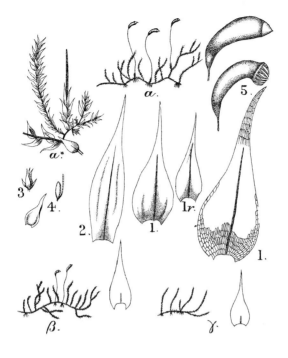

FIGURE 112. *Amblystegiu serpens*
(after Braithwaite).

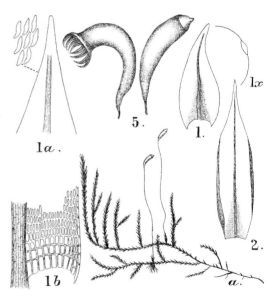

FIGURE 113. *Hygroamblystegium fluviatile*
(after Braithwaite).

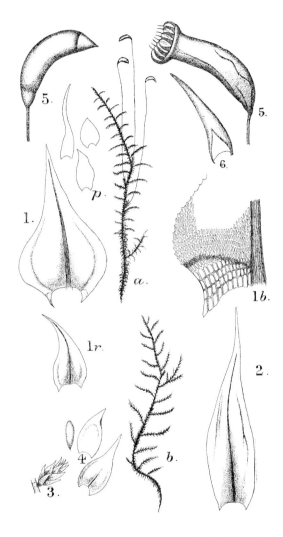

FIGURE 114. *Cratoneuron filicinum* (after Braithwaite).

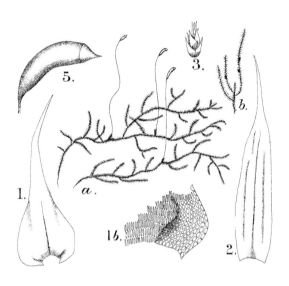

FIGURE 115. *Campylium chrysophyllum* (after Braithwaite).

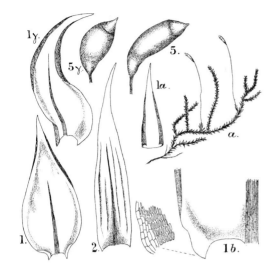

FIGURE 116. *Hygrohypnum luridum* (after Braithwaite).

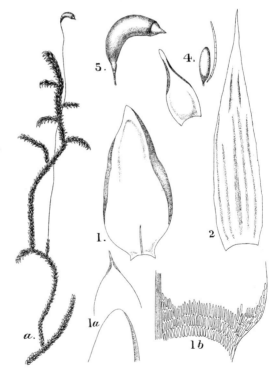

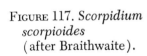

FIGURE 117. *Scorpidium scorpioides* (after Braithwaite).

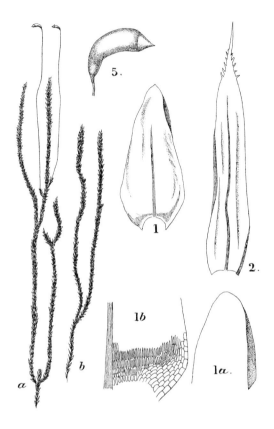

FIGURE 118. *Calliergon stramineum* (after Braithwaite).

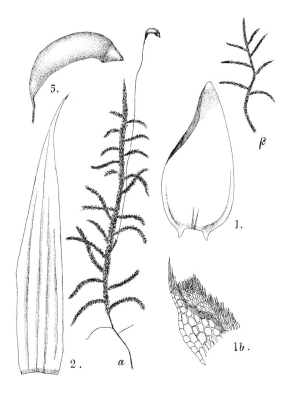

FIGURE 119. *Calliergonella cuspidata* (after Braithwaite).

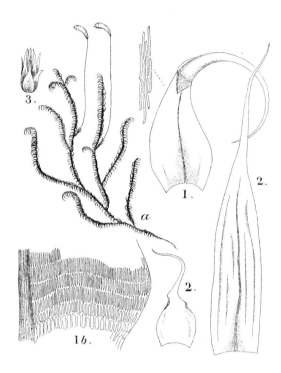

FIGURE 120. *Drepanocladus revolvens* (after Braithwaite).

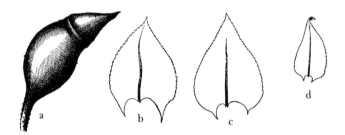

FIGURE 121. *Bryhnia novae-angliae* (after Limpricht).

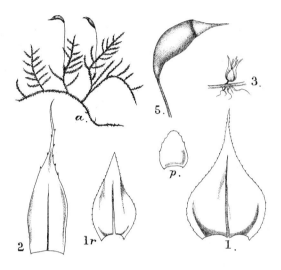

FIGURE 122. *Eurhynchium pulchellum* (after Braithwaite).

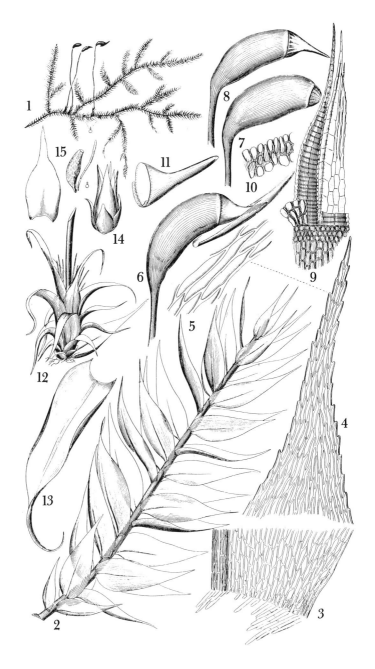

FIGURE 123. *Rhynchostegium serrulatum* (after Sullivant).

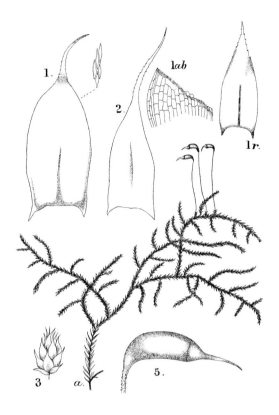

FIGURE 124. *Cirriphyllum piliferum* (after Braithwaite).

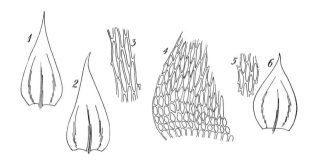

FIGURE 125. *Chamberlainia cyrtophylla* (after Grout).

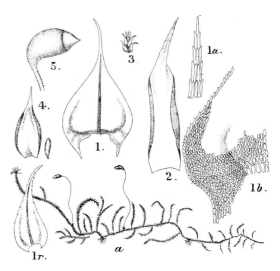

FIGURE 126. *Brachythecium reflexum* (after Braithwaite).

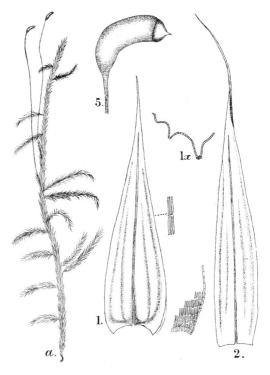

FIGURE 127. *Camptothecium nitens* (after Braithwaite).

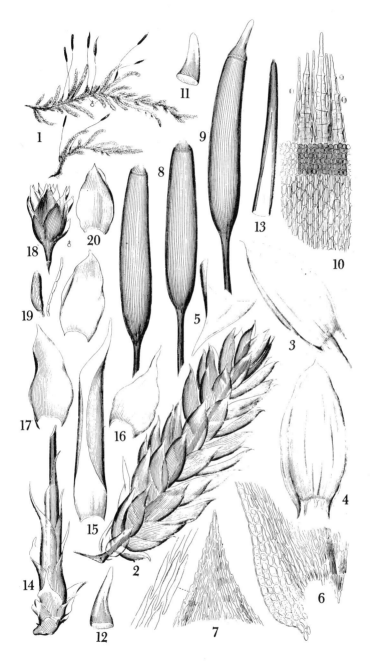

FIGURE 128. *Entodon seductrix* (after Sullivant).

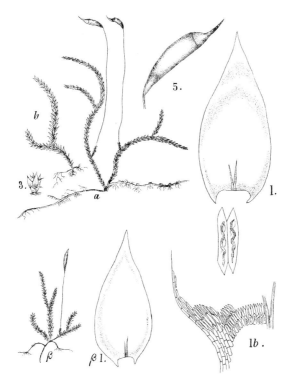

FIGURE 129. *Plagiotheciu* *sylvaticum* (after Braithwaite).

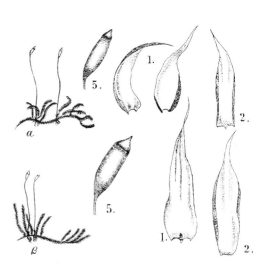

FIGURE 130. *Isopterygium pulchellum* (after Braithwaite).

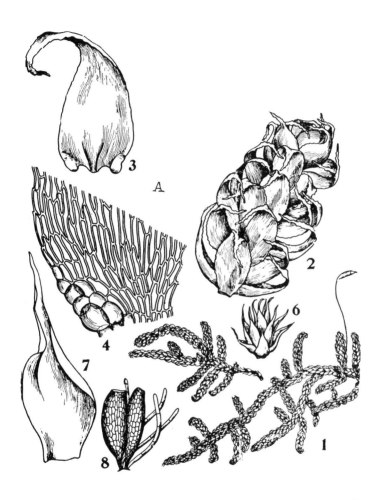

FIGURE 131. *Brotherella recurvans* (after Britton).

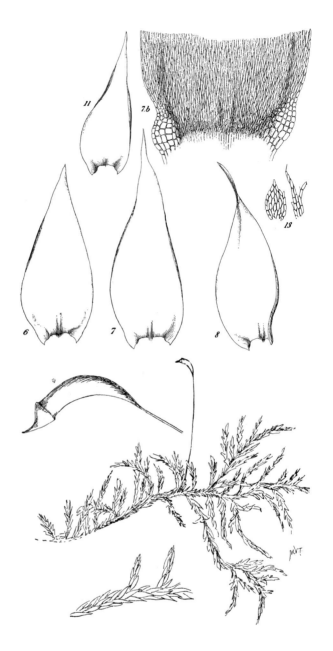

FIGURE 132. *Heterophyllium haldanianum* (after *Bryologia Europæa*).

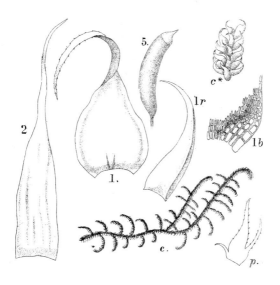

FIGURE 133. *Hypnum imponens* (after Braithwaite).

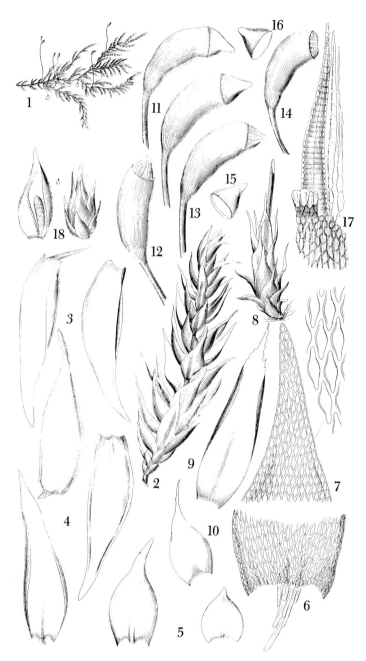

FIGURE 134. *Homomallium adnatum* (after Sullivant).

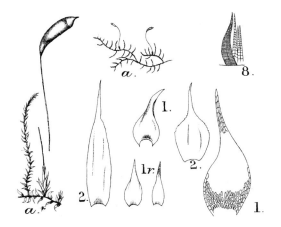

FIGURE 135. *Amblystegiella confervoides* (after Braithwaite).

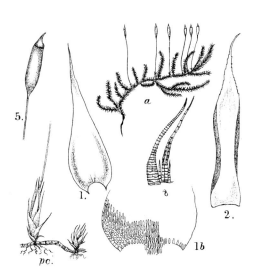

FIGURE 136. *Pylaisia polyantha* (after Braithwaite).

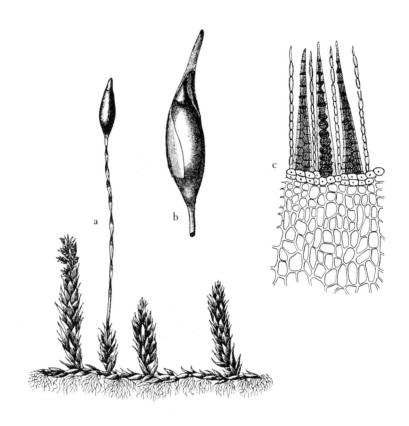

FIGURE 137. *Platygyrium repens* (after Limpricht).

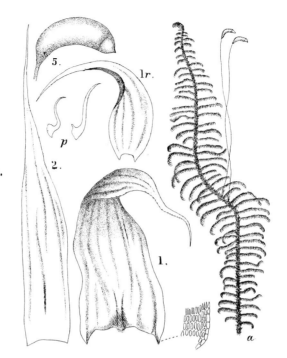

FIGURE 138. *Ptilium crista-castrensis* (after Braithwaite).

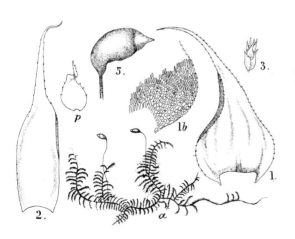

FIGURE 139. *Ctenidium molluscum* (after Braithwaite)

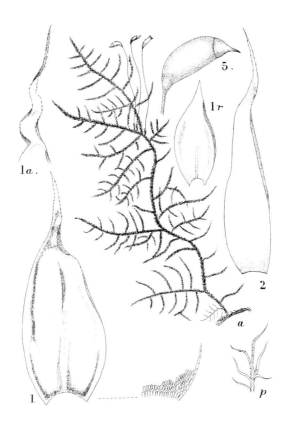

FIGURE 140. *Hylocomium splendens* (after Braithwaite).

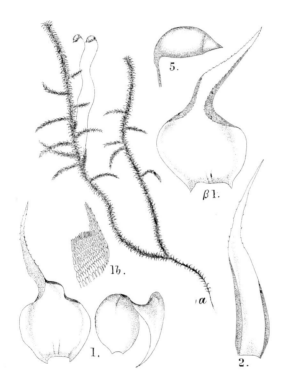

FIGURE 141. *Rhytidiadelphus squarrosus* (after Braithwaite).

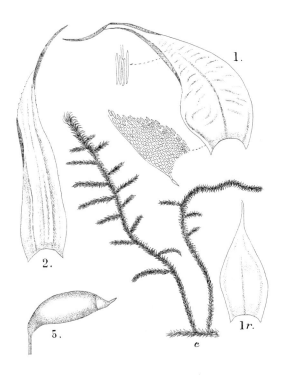

FIGURE 142. *Rhytidium rugosum* (after Braithwaite).

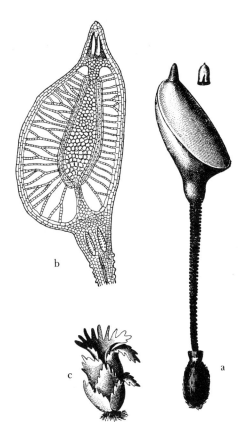

FIGURE 143. *Buxbaumia aphylla*
(after Limpricht).

b

c

a

FIGURE 144. *Diphyscium foliosum*
(after Limpricht).

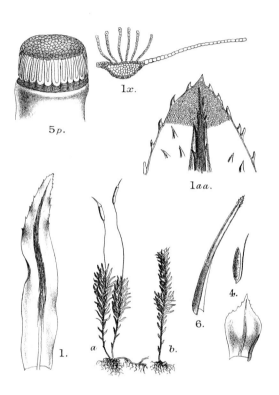

FIGURE 145. *Atrichum angustatum* (after Braithwaite).

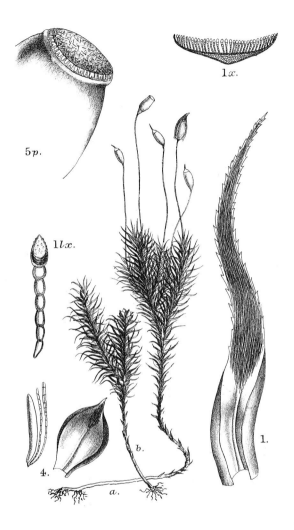

FIGURE 146. *Pogonatum alpinum* (after Braithwaite).

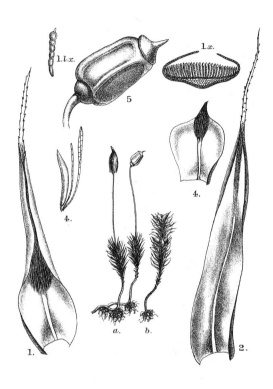

FIGURE 147. *Polytrichum piliferum* (after Braithwaite).

3. *Pohlia* Hedw.

Plants in loose or dense tufts; leaves mostly lanceolate, not bordered, often larger and more crowded above; costa ending below
the apex to excurrent; cells narrow and elongate; capsules inclined
to pendulous, shortly ovoid to clavate or cylindric, usually with a
distinct neck; peristome double, endostome with a basal membrane,
segments, and usually cilia.

1. *Pohlia cruda* (Hedw.) Lindb. — Light-green, lustrous plants
in loose tufts; leaves lanceolate, acuminate, larger toward the stem
tip; cells very long and narrow; spores in May to June. — On rocks
or soil, particularly in damp, shaded niches of banks or cliffs. Across
the continent, southward to Pennsylvania in the East and Colorado,
Arizona, and California in the West; nearly cosmopolitan.

ILLUSTRATIONS: Figure 74. Grout, *Moss Flora of North America*, vol. 2, Pl.
74A.

Upper Peninsula: Alger Co., *Steere*. Keweenaw Co., *Cooper*. Marquette Co.,
Nichols. Ontonagon Co., *Steere*.

Lower Peninsula: Alpena Co., *Hall*. Cheboygan Co., *Nichols*. Emmet Co.,
Steere.

2. *Pohlia acuminata* Hoppe & Hornsch. — Plants dull-green,
loosely tufted, rarely more than 1 cm. high, radiculose below; leaves

lance-acuminate, slightly denticulate at the apex; upper cells linear, thick-walled; capsules suberect to subpendulous, clavate to subcylindric, with a neck shorter to nearly as long as the urn; cilia none; spores in late summer to early autumn. — In crevices of rocks; Canada and the northern United States; Europe and Asia.

ILLUSTRATION: Grout, *Moss Flora of North America*, vol. 2, Pl. 75B.

Upper Peninsula: Alger Co. (Miner's Falls), *Steere.*

3. **Pohlia elongata** Hedw. — Similar to *P. acuminata;* leaf cells long-rhomboidal, thin-walled; neck of capsule as long as the urn or longer. — On soil in woods; across southern Canada, south to North Carolina, the Great Lakes, and British Columbia; Europe, North Africa, and Asia.

ILLUSTRATIONS: Grout, *Mosses with Hand-Lens and Microscope*, fig. 106, *Moss Flora of North America*, vol. 2, Pl. 75C.

Upper Peninsula: Alger Co. (near Munising), *Nichols.*

4. **Pohlia nutans** (Hedw.) Lindb. — Plants dull-green or yellowish, usually about 1–2 cm. high; leaves narrowly to broadly lance-acuminate, slightly denticulate above; cells rhomboidal to linear, thick-walled; capsules elongate-pyriform, the neck not particularly evident, shorter than the urn; cilia 2–3, nodose or ± appendiculate; spores in early summer. — Very common on various substrata, especially on rotten stumps. Across the continent, south at high elevations to North Carolina, Colorado, Arizona, and California; nearly cosmopolitan.

ILLUSTRATIONS: Conard, *How to Know the Mosses* (ed. 2), fig. 145a–d. Grout, *Mosses with Hand-Lens and Microscope*, fig. 107. Jennings, *Mosses of Western Pennsylvania* (ed. 2), Pl. 21. Welch, *Mosses of Indiana*, fig. 117.

Upper Peninsula: Alger Co., *Gilly & Parmelee.* Chippewa Co., *Thorpe.* Gogebic Co., *Bessey.* Houghton Co., *Parmelee.* Keweenaw and Mackinac counties, *Gilly & Parmelee.* Marquette Co., *Nichols.* Ontonagon Co., *Nichols & Steere.*
Lower Peninsula: Alpena Co., *Robinson & Wells.* Barry Co., *Gilly & Parmelee.* Cheboygan Co., *Nichols.* Crawford Co., *Gilly & Parmelee.* Eaton Co., *Darlington.* Emmet Co., *Gilly & Parmelee.* Genesee, Ingham, Lapeer, and Leelanau counties, *Darlington.* Macomb Co., *Cooley.* Washtenaw Co., *Kauffman.*

5. **Pohlia proligera** (Limpr.) Lindb. — Small plants, yellow-green, glossy; vermicular, twisted and contorted brood-bodies abundant in leaf axils; leaves ovate-lanceolate; cells linear, thin-walled; generally sterile. — On soil of banks and crevices of cliffs. Greenland; across the continent, south to New York, the Great Lakes, and Colorado; Europe, Macaronesia, and Asia.

ILLUSTRATIONS: Grout, *Mosses with Hand-Lens and Microscope*, fig. 109, *Moss Flora of North America*, vol. 2, Pl. 78.

Upper Peninsula: Alger Co., *Nichols*. Keweenaw Co., *Steere*. Luce Co., *Nichols*. Ontonagon Co., *Nichols & Steere*.

6. *Pohlia bulbifera* (Warnst.) Warnst. — Small plants (up to about 1.5 cm. high), yellow-green, shiny; brood-bodies 1–3, ovoid, in the leaf axils; leaves lanceolate, acute to slightly acuminate, strongly toothed; cells rhomboid-vermicular, thin-walled; fruit uncommon. — On soil, particularly on banks. Greenland; Nova Scotia to Massachusetts and Michigan; Europe, Azores, and Asia.

ILLUSTRATION: Grout, *Moss Flora of North America*, vol. 2, Pl. 78.

Upper Peninsula: Luce Co. (Whitefish Point), *Griffin*.

7. *Pohlia pulchella* (Hedw.) Lindb. — Plants small (up to 1 cm. high), dull-green; leaves lanceolate, slightly toothed above; cells long-linear, thin-walled; setae flexuose and geniculate; capsules short, ovoid-pyriform, light-brown; stomata superficial; annulus present, usually adhering to the operculum; peristome teeth yellow or light-brown; cilia of endostome well developed; spores in May. — On moist, clayey soil. Eastern North America from Nova Scotia to Washington, D.C., and Michigan; Europe and Asia.

ILLUSTRATIONS: Grout, *Mosses with Hand-Lens and Microscope*, Pl. 45 (as *P. lescuriana*). Jennings, *Mosses of Western Pennsylvania* (ed. 2), Pl. 22.

Upper Peninsula: Mackinac Co., *Nichols*.
Lower Peninsula: Cheboygan Co., *Nichols*.

8. *Pohlia wahlenbergii* (Web. & Mohr) Andr. — Whitish-green plants with reddish stems, in loose tufts 2–6 cm. high; leaves ovate-lanceolate, denticulate above; cells rhomboidal, lax and thin-walled; setae flexuose, up to 2 or more cm. long; capsules ovoid-pyriform, light- to dark-brown; annulus none; stomata immersed; peristome teeth yellow-brown; cilia of endostome well developed; spores in late spring (fruits uncommon). — On wet soil. Nearly cosmopolitan.

ILLUSTRATIONS: Conard, *How to Know the Mosses* (ed. 2), fig. 144. Grout, *Mosses with Hand-Lens and Microscope*, Pl. 46 (as *Mniobryum albicans*). Jennings, *Mosses of Western Pennsylvania* (ed. 2), Pl. 22 (as *Mniobryum*). Welch, *Mosses of Indiana*, fig. 118.

Upper Peninsula: Keweenaw Co., *Allen & Stuntz*. Ontonagon Co., *Darlington & Bessey*.
Lower Peninsula: Cheboygan Co., *Nichols*. Eaton and Genesee counties, *Darlington*. Kalamazoo Co., *Becker*. Leelanau Co., *Darlington*.

9. *Pohlia filiformis* (Dicks.) Andr. — Plants small, up to 1 cm. high, terete, yellow-green, shiny; leaves imbricate, oval or broadly oblong-ovate, obtuse; margins plane, entire or slightly crenulate above; cells rhomboidal to vermicular; capsules clavate or long-cylindric; cilia nodose or ± appendiculate. — On soil among rocks or on cliffs. Greenland; Canada and Alaska to New York and Minnesota; Europe, Africa, Central and South America.

ILLUSTRATION: Grout, *Moss Flora of North America,* vol. 2, Pl. 83c.

Upper Peninsula: Alger Co., *Steere.* Gogebic Co., *Conard.* Keweenaw Co., *Steere.*

4. *Bryum* Hedw.

Plants usually in dense, mostly dull tufts; leaves lanceolate or ovate, often acuminate or ± awned; costa usually excurrent; cells oblong-hexagonal, relatively large; capsules pendulous, usually with a fairly well-marked neck; peristome double, the endostome with a high basal membrane, segments and cilia usually well developed.

1. Species of poor or dry soil, common and weedy,
   in exposed places 2
1. Species of moist, generally shady places, not weedy 3
   2. Plants silvery-white; leaves ovate; costa
      ending below the apex . . . . 12. *B. argenteum*
   2. Plants green or yellowish-green; leaves ovate-
      lanceolate; costa excurrent . . . . . 9. *B. caespiticium*
3. Endostome with cilia rudimentary or lacking 4
3. Endostome with well-developed cilia 6
   4. Lamellae on the inner side of the peristome teeth
      freely joined by vertical bars . . . . 1. *B. angustirete*
   4. Lamellae not joined 5
5. Leaves strongly bordered; cells not particularly lax;
   synoicous; capsules about 2.5 mm. long . . . 2. *B. inclinatum*
5. Leaves not strongly bordered; cells lax; autoicous;
   capsules up to 6 mm. long . . . . . . 3. *B. uliginosum*
   6. Leaves obtuse or rounded at apex 7
   6. Leaves acute or acuminate 8
7. Leaf cells thin-walled . . . . . . 6. *B. tortifolium*
7. Leaf cells with moderately thick walls . . 11. *B. muehlenbeckii*
   8. Outer peristome light-yellow 9
   8. Outer peristome dark-yellow or brownish
      (at least below) 10
9. Leaves long-decurrent, not strongly bordered . . 5. *B. weigelii*
9. Leaves not particularly decurrent, strongly bordered . 4. *B. pallens*
   10. Capsules cylindric, the neck short and
       rounded . . . . 14. *B. coronatum*
   10. Capsules elongate, the neck tapered to the seta 11
11. Costa percurrent or nearly so 12
11. Costa excurrent 13

[88]

12. Leaves not particularly decurrent; cells lax
   and thin-walled . . . . 13. *B. capillare*
12. Leaves conspicuously decurrent; cells
   not lax . . . . 10. *B. pseudotriquetrum*
13. Autoicous plants . . . . 8. *B. pallescens*
13. Synoicous plants . . . . 7. *B. cuspidatum*

1. *Bryum angustirete* Kindb. — Plants closely tufted, with reddish, branched stems, generally about 5 mm. high; leaves ovate-lanceolate, acuminate; costa long-excurrent as a slightly denticulate point; cells oblong-hexagonal, with thickened corners; synoicous; peristome brown, the lamellae on the inner surface extensively joined by vertical bars, endostome partially adhering to the teeth, cilia none or rudimentary; spores in summer. — On soil, often in rock crevices. Greenland; arctic America south to Washington, D.C., Missouri, Colorado, and Arizona; Europe and Asia.

ILLUSTRATIONS: Conard, *How to Know the Mosses* (ed. 2), fig. 148e (as *B. pendulum*). Grout, *Mosses with Hand-Lens and Microscope*, fig. 110 (as *B. pendulum*). Jennings, *Mosses of Western Pennsylvania* (ed. 2), Pl. 22. Welch, *Mosses of Indiana*, fig. 119 (as *B. pendulum*).

Upper Peninsula: Keweenaw Co., *Cooper.*
Lower Peninsula: Emmet Co., *Nichols.*

2. *Bryum inclinatum* (Web. & Mohr) Sturm — Plants in dense, green or yellow-green tufts, up to 1 cm. or more high; leaves ovate-lanceolate; costa long-excurrent as an entire or slightly denticulate point; cells oblong-hexagonal, moderately thick-walled; synoicous; peristome teeth reddish at insertion, yellow above, endostome free, cilia rudimentary; spores in summer. — On soil. Greenland; arctic America south to Maine, Michigan, Colorado, and California; Europe, North Africa, Asia, and southern South America.

ILLUSTRATION: Grout, *Moss Flora of North America*, vol. 2, Pl. 85.

Upper Peninsula: Keweenaw Co. (Isle Royale), *Allen & Stuntz.*

3. *Bryum uliginosum* (Brid.) BSG — Plants loosely tufted, rarely more than 1 cm. high; leaves ovate-lanceolate, acuminate; costa ending slightly below the apex to short-excurrent; cells oblong-hexagonal, rather large and thin-walled; autoicous; capsules generally very long and slender (up to 6 mm. long), slightly curved; exostome free, cilia short or lacking; spores in summer. — On wet soil. Newfoundland to the Yukon, south to Ohio and New Mexico; Europe and Asia.

ILLUSTRATIONS: Conard, *How to Know the Mosses* (ed. 2), fig. 148a–d. Grout, *Moss Flora of North America*, vol. 2, Pl. 87 (as *B. cernuum*).

Lower Peninsula: Cheboygan Co., *Nichols.* Kalamazoo Co., *Becker.* Mecosta Co., *Schnooberger.* Presque Isle Co., *Steere.*

4. *Bryum pallens* Sw. — Plants in loose tufts about 1 cm. high; leaves obovate, gradually acuminate, entire; costa percurrent; cells rather large, thin-walled; dioicous; capsules about 5.5 mm. long; exostome yellow; cilia of endostome variable, lacking to well developed and appendiculate; spores 18–22 μ, maturing in summer. — On soil or rocks in wet places. Greenland; across the continent, south to New York and Washington; Europe and Asia.

ILLUSTRATION: Grout, *Mosses with Hand-Lens and Microscope*, fig. 112.

Upper Peninsula: Alger Co., *Gleason.* Keweenaw Co., *Holt.* Marquette Co., *Nichols.*
Lower Peninsula: Cheboygan Co., *Nichols.*

5. *Bryum weigelii* Spreng. — Plants in loose, light-green tufts up to 3 cm. or more in height; leaves ovate or ovate-lanceolate, acute or acuminate, long-decurrent; margins entire, plane or reflexed only below; costa percurrent or nearly so; cells oblong-hexagonal, thin-walled; dioicous; capsules about 3.5 mm. long; cilia appendiculate; spores 10–15 μ, mature in summer. — In wet places. Greenland; across the continent, south to New York in the East, Colorado in the West; Europe and Asia.

ILLUSTRATIONS: Grout, *Mosses with Hand-Lens and Microscope*, fig. 111 (as *B. duvalii*), *Moss Flora of North America*, vol. 2, Pl. 89B.

Upper Peninsula: Chippewa Co., *Steere.* Marquette Co., *Nichols.*
Lower Peninsula: Cheboygan Co., *Nichols.*

6. *Bryum tortifolium* Funck — Plants in low, soft tufts; leaves soft, ovate, rounded-obtuse; margins entire, plane; costa ending below the apex; cells broadly hexagonal, thin-walled; dioicous; fruits rare; cilia appendiculate. — On moist, sandy soil. Greenland; arctic America south to Pennsylvania, the Great Lakes, and British Columbia; Europe and Asia.

ILLUSTRATION: Grout, *Moss Flora of North America*, vol. 2, Pl. 89A.

Upper Peninsula: Chippewa Co. (Sugar Island), *Steere.*

7. *Bryum cuspidatum* (BSG) Schimp. — Plants in dense tufts up to 1 cm. high; leaves ovate-lanceolate, acuminate, crowded in a rosette-like tuft; costa excurrent; cells oblong-hexagonal, relatively long above; synoicous; cilia well developed, appendiculate; spores about 14 μ, mature in summer. — On moist soil, rocks, or rotten wood. Throughout North America; Europe and Asia.

[90]

ILLUSTRATIONS: Grout, *Moss Flora of North America*, vol. 2, Pl. 92B. Jennings, *Mosses of Western Pennsylvania* (ed. 2), Pl. 23. Welch, *Mosses of Indiana*, fig. 120.

Upper Peninsula: Chippewa Co., *Steere.* Keweenaw Co., *Allen & Stuntz.* Marquette Co., *Nichols.* Ontonagon Co., *Darlington.*
Lower Peninsula: Ingham Co., *Wheeler.* Montcalm Co., *Cooley.* Van Buren Co., *Kauffman.* Washtenaw Co., *Steere.*

8. **Bryum pallescens** Schleich. — Closely related to *B. cuspidatum;* leaves spirally twisted when dry; costa excurrent in uppermost leaves; autoicous; spores about 20 μ, ripe in May or June. — On moist soil or rock. Greenland; arctic America, south to Ohio in the East, Arizona in the West; Europe, Africa, and Asia.

ILLUSTRATIONS: Grout, *Moss Flora of North America*, vol. 2, Pl. 90B. Jennings, *Mosses of Western Pennsylvania* (ed. 2), Pl. 23.

Lower Peninsula: Emmet Co., *Nichols.*

9. **Bryum caespiticium** Hedw. — Plants densely tufted, rarely more than 1 cm. high; leaves ovate-lanceolate, long-acuminate, crowded in a rosette; costa long-excurrent; upper cells rhomboidal, relatively long and narrow; dioicous; cilia appendiculate; spores 7–10 μ, maturing in May or June. — On poor soil; weedy. Nearly cosmopolitan.

ILLUSTRATIONS: Conard, *How to Know the Mosses* (ed. 2), fig. 151a–d. Grout, *Mosses with Hand-Lens and Microscope*, fig. 113. Jennings, *Mosses of Western Pennsylvania* (ed. 2), Pl. 24. Welch, *Mosses of Indiana*, fig. 121.

Upper Peninsula: Alger and Chippewa counties, *Steere.* Keweenaw Co., *Cooper.* Ontonagon Co., *Darlington.*
Lower Peninsula: Alpena Co., *Wheeler.* Cheboygan Co., *Nichols.* Eaton and Ingham counties, *Wheeler.* Leelanau Co., *Darlington.* Macomb Co., *Cooley.* Washtenaw Co., *Kauffman.*

10. **Bryum pseudotriquetrum** (Hedw.) Schwaegr. — Plants in loose tufts up to 3 cm. or more in height, dark-green with red stems, brownish-tomentose below; leaves rather widely spaced, oblong-lanceolate, gradually acuminate, long-decurrent; costa percurrent or slightly excurrent; cells oblong-hexagonal, moderately incrassate and pitted; synoicous or dioicous; cilia appendiculate; spores about 14 μ, maturing in summer. — On soil in wet places. Greenland; especially in arctic and boreal America but also found almost throughout the United States; Europe and Asia.

ILLUSTRATIONS: Figure 75. Grout, *Mosses with Hand-Lens and Microscope*, Pl. 47 (as *B. bimum*). Jennings, *Mosses of Western Pennsylvania* (ed. 2), Pl. 23. Welch, *Mosses of Indiana*, fig. 122.

Upper Peninsula: Alger Co., *Wheeler*. Chippewa Co., *Steere*. Keweenaw Co., *Allen* & *Stuntz*. Ontonagon Co., *Darlington* & *Bessey*.

Lower Peninsula: Cheboygan Co., *Nichols*. Clare Co., *Schnooberger*. Emmet Co., *Nichols*. Leelanau Co., *Darlington*. Montcalm Co., *Schnooberger*. Washtenaw Co., *Steere*.

11. *Bryum muehlenbeckii* BSG — Plants in dense tufts up to 3 cm. high, red- or brown-tinted; leaves crowded and imbricate, oblong-ovate, obtuse or subacute; margins entire, somewhat reflexed; costa percurrent or nearly so; cells rhomboid-hexagonal, moderately thick-walled and pitted; dioicous; cilia slightly appendiculate; spores 15–18 μ, maturing in summer. — On wet rocks. Greenland; southern Canada and the northern United States; Europe and Asia.

ILLUSTRATION: Grout, *Moss Flora of North America*, vol. 2, Pl. 96A.

Upper Peninsula: Keweenaw Co., *Cooper*. Marquette Co., *Nichols*. Ontonagon Co., *Nichols* & *Steere*.

12. *Bryum argenteum* Hedw. — Small, silvery plants with crowded and imbricate, concave, broadly ovate, cuspidate or acuminate leaves; costa ending below the apex; basal cells quadrate or short-rectangular, green; upper cells oblong-hexagonal or rhomboidal, pale; dioicous; capsules short, usually red; cilia appendiculate; spores 14–18 μ, maturing in winter or early spring. — A common weed on dry, sandy soil, often on paths or in cracks of sidewalks. Cosmopolitan.

ILLUSTRATIONS: Conard, *How to Know the Mosses* (ed. 2), fig. 149. Grout, *Mosses with Hand-Lens and Microscope*, fig. 115. Jennings, *Mosses of Western Pennsylvania* (ed. 2), Pl. 24.

Upper Peninsula: Baraga Co., *Parmelee*. Keweenaw Co., *Povah*. Ontonagon Co., *Darlington*.

Lower Peninsula: Cheboygan Co., *Nichols*. Clinton Co., *Parmelee*. Ingham and Leelanau counties, *Darlington*. Washtenaw Co., *Kauffman*.

12a. *Bryum argenteum* var. *lanatum* (P.-Beauv.) BSG — Leaf tips longer; costa percurrent or slightly excurrent. — Ranging with the species.

Upper Peninsula: Keweenaw Co., *Thorpe; Povah*.

13. *Bryum capillare* Hedw. — Low plants (less than 1 cm. high), soft, dark-green or brown, usually densely radiculose below; dark-yellow, filiform propagula often abundant on stems; leaves obovate or oblong-lanceolate, cuspidate or acuminate, soft, spirally twisted when dry; margins ± reflexed below; costa ending below the apex

or sometimes excurrent; dioicous; cilia appendiculate; spores 10–16 μ, maturing in summer. — On moist humus, bark of trees, decayed wood, etc. Nearly cosmopolitan.

ILLUSTRATIONS: Conard, *How to Know the Mosses* (ed. 2), fig. 151e. Grout, *Mosses with Hand-Lens and Microscope,* Pl. 48. Jennings, *Mosses of Western Pennsylvania* (ed. 2), Pl. 24. Welch, *Mosses of Indiana,* fig. 125.

Upper Peninsula: Keweenaw Co., *Cooper.* Marquette Co., *Nichols.* Ontonagon Co., *Nichols* & *Steere.*

Lower Peninsula: Alpena Co., *Robinson* & *Wells.* Cheboygan Co., *Nichols.* Leelanau Co., *Darlington.* Van Buren Co., *Kauffman.*

14. *Bryum coronatum* Schwaegr. — Plants low (rarely 1.5 cm. high), rather densely tufted; leaves narrowly ovate, sharply acuminate; margins entire, revolute; costa percurrent to long-excurrent; cells narrowly oblong-hexagonal; dioicous; capsules dark-red when mature, cylindric with a short, rounded neck wider than the urn when dry; cilia appendiculate; spores about 10 μ. — On soil or rock. Florida; pantropical (probably adventive in Michigan, where it has been collected only once, in a lime sink).

ILLUSTRATION: Grout, *Moss Flora of North America,* vol. 2, Pl. 92A.

Lower Peninsula: Alpena Co., *Chopra* & *Sharp.*

5. *Rhodobryum* (Schimp.) Hampe

Plants robust, loosely tufted; stems erect from underground "rhizomes"; leaves crowded in terminal rosettes, smaller and scale-like below, obovate, acuminate, generally bordered and toothed; costa strong; cells oblong-hexagonal, relatively large and broad; setae generally clustered; capsules cylindric, short-necked, horizontal or pendulous; peristome double, cilia nodulose or appendiculate.

Rhodobryum roseum (Hedw.) Limpr. — Plants dark-green, up to 3 or 4 cm. high; leaves large, crowded in conspicuous rosettes; margins serrate above, revolute below; setae generally clustered; spores 20–25 μ, maturing in late autumn or winter. — On soil, rocks, logs, and bark at base of trees. Widespread in eastern North America, less common in the West; Mexico; Europe and Asia.

ILLUSTRATIONS: Figure 76. Conard, *How to Know the Mosses* (ed. 2), fig. 147. Jennings, *Mosses of Western Pennsylvania* (ed. 2), Pl. 24.

Upper Peninsula: Alger Co., *Wheeler.* Chippewa Co., *Steere.* Houghton Co., *Parmelee.* Keweenaw Co., *Cooper.* Mackinac Co., *Wynne.* Ontonagon Co., *Nichols* & *Steere.*

Lower Peninsula: Alpena Co., *Robinson* & *Wells.* Barry Co., *Gilly* & *Parmelee.* Cheboygan Co., *Nichols.* Clinton Co., *Parmelee.* Leelanau Co., *Darlington.*

MNIACEAE

Loosely tufted, relatively robust and conspicuous plants; stems normally erect; leaves broad, often arranged in terminal rosettes, mostly bordered and toothed; costa strong, single; cells hexagonal or subquadrate, smooth or rarely mammillose; inflorescences terminal; setae elongate, sometimes clustered; capsules cylindric, horizontal to pendent; peristome double; calyptra cucullate.

Inner peristome about as long as the outer, divided into
  segments and cilia . . . . 1. *Mnium*
Inner peristome longer than the outer, coalesced into
  a dome-like structure supported by segments alternating
  with the exostome teeth . . . . 2. *Cinclidium*

1. *Mnium* Hedw.

Leaves broadly oblong to lingulate, rounded-obtuse to cuspidate or acuminate, mostly bordered by linear cells, entire or singly or doubly serrate; costa ending below the apex to short-excurrent; cells small; inner peristome about the same length as the outer, consisting of free segments and cilia from a basal membrance.

1. Leaves entire (or nearly so) 2
1. Leaves serrate 2
  2. Leaves indistinctly bordered, rarely sparsely
    and obscurely toothed; cells elongate, in
    oblique rows . . . . 10. *M. cinclidioides*
  2. Leaves distinctly bordered, entire; cells not
    elongate or obliquely arranged 3
3. Dioicous 4
3. Synoicous . . . . 12. *M. pseudopunctatum*
  4. Leaves very small, rounded-obtuse; costa ending
    well below the apex . . . . . 13. *M. andrewsianum*
  4. Leaves rather large; costa percurrent or excurrent
    into a bluntly cuspidate apex . . . . 11. *M. punctatum*
5. Leaves not bordered; costa ending well below
  the apex . . . . 1. *M. stellare*
5. Leaves clearly bordered; costa percurrent or excurrent . . . 6
  6. Leaves doubly serrate 7
  6. Leaves singly serrate 10
7. Synoicous (or paroicous) 8
7. Dioicous 9
  8. Leaf cells mostly more than 25 μ, irregularly
    quadrate with thickened corners, in
    longitudinal rows . . . . 4. *M. serratum*
  8. Leaf cells mostly less than 25 μ, hexagonal, evenly
    thin-walled, not in obvious rows . . . 5. *M. spinulosum*
9. Leaf cells small (rarely more than 20 μ), not
  thickened at the corners . . . . . 2. *M. orthorhynchum*
9. Leaf cells more than 20 μ, with thickened
  corners . . . . 3. *M. lycopodioides*

10. Leaf cells hexagonal, with straight,
    thin walls . . . . 6. *M. drummondii*
10. Leaf cells rounded or oval, with thickened corners 11
11. Leaves toothed only in the upper half; cells
    less than 25 μ, rounded . . . . 7. *M. cuspidatum*
11. Leaves toothed nearly to the base 12
    12. Synoicous; marginal teeth 1-celled . . . 8. *M. medium*
    12. Dioicous; marginal teeth 1–3-celled . . . . 9. *M. affine*

1. *Mnium stellare* Hedw. — Plants dark-green, often becoming reddish; leaves oblong-elliptic, obtuse or short-pointed, irregularly and sometimes ± doubly serrate, unbordered; costa ending well below the apex; cells thickened at the corners; dioicous; spores maturing in May or June. — On humus or bark at base of trees. Widespread in eastern North America; Europe and Asia.

ILLUSTRATIONS: Conard, *How to Know the Mosses* (ed. 2), fig. 157. Grout, *Mosses with Hand-Lens and Microscope*, fig. 121. Jennings, *Mosses of Western Pennsylvania* (ed. 2), Pl. 26. Welch, *Mosses of Indiana*, fig. 127.

Upper Peninsula: Chippewa Co., *Steere*. Houghton Co., *Parmelee*. Marquette Co., *Nichols*. Ontonagon Co., *Darlington & Bessey*.
Lower Peninsula: Alpena Co., *Robinson & Wells*. Cheboygan Co., *Nichols*. Leelanau Co., *Darlington*. Washtenaw Co., *Steere*.

2. *Mnium orthorhynchum* Brid. — Leaves oblong-ovate, acute or apiculate, bordered, doubly serrate; costa percurrent; cells up to 20 μ, hexagonal, thin-walled, not thickened at the corners; dioicous; spores ripening in summer. — On soil and rocks. Greenland; across the continent, south to North Carolina and New Mexico; Europe, North Africa, and Asia.

ILLUSTRATIONS: Conard, *How to Know the Mosses* (ed. 2), fig. 156a–c. Grout, *Mosses with Hand-Lens and Microscope*, Pl. 51.

Upper Peninsula: Keweenaw Co., *Cooper*. Marquette Co., *Nichols*. Ontonagon Co., *Nichols & Steere*.
Lower Peninsula: Alpena Co., *Robinson & Wells*. Cheboygan Co., *Nichols*. Eaton Co., *Darlington*. Van Buren Co., *Kauffman*.

3. *Mnium lycopodiodes* Schwaegr. — Leaves ovate-lanceolate, or acuminate, bordered, doubly serrate; costa percurrent or excurrent; cells 20–30 μ (or more), rounded-hexagonal, delicate and thin-walled with thickened corners; dioicous; spores maturing in July. — On rocks beside brooks. Southern Canada and northern United States; Europe and Asia.

ILLUSTRATION: Grout, *Moss Flora of North America*, vol. 2, Pl. 104.

Upper Peninsula: Alger Co., Steere.
Lower Peninsula: Cheboygan Co., *Steere*.

4. *Mnium serratum* Brid. — Leaves oblong-ovate, short-acuminate, bordered, doubly serrate; costa percurrent; cells up to 35 μ, irregularly quadrate, thickened at the corners, arranged in ± longitudinal rows; synoicous; spores maturing in May. — On soil or rocks. Across the continent, south to Tennessee, Missouri, and Arizona; Europe and Asia.

ILLUSTRATIONS: Figure 77. Conard, *How to Know the Mosses* (ed. 2), fig. 156d. Grout, *Mosses with Hand-Lens and Microscope*, Pl. 51. Jennings, *Mosses of Western Pennsylvania* (ed. 2), Pl. 25. Welch, *Mosses of Indiana*, fig. 129.

Upper Peninsula: Keweenaw Co., *Cooper.* Menominee Co., *Hill.*
Lower Peninsula: Alpena Co., *Robinson & Wells.* Cheboygan Co., *Nichols.* Ingham and Leelanau counties, *Darlington.*

5. *Mnium spinulosum* (Voit) Schwaegr. — Leaves broadly obovate, short-cuspidate, bordered, doubly serrate; costa slightly excurrent; cells 18–25 μ, irregularly hexagonal, evenly thin-walled; synoicous (or paroicous); spores in May to June. — On soil in woods (especially in coniferous woods). Across the continent, south to Maryland and Oregon; Europe.

ILLUSTRATIONS: Conard, *How to Know the Mosses* (ed. 2), fig. 156e. Grout, *Mosses with Hand-Lens and Microscope*, fig. 118f–h. Jennings, *Mosses of Western Pennsylvania* (ed. 2), Pl. 65.

Upper Peninsula: Chippewa Co., *Thorpe.* Keweenaw Co., *Parmelee.* Luce and Marquette counties, *Nichols.* Ontonagon Co., *Darlington & Bessey.*
Lower Peninsula: Cheboygan and Emmet counties, *Nichols.* Leelanau and Oakland counties, *Darlington.*

6. *Mnium drummondii* Bruch & Schimp. — Sterile stems stoloniform; fertile stems erect; leaves broadly obovate, short-acuminate, bordered, singly serrate in the upper part only; costa percurrent; cells regularly hexagonal, with thin walls, up to 40 μ; synoicous (but frequently producing antheridial heads also); sporophytes often clustered; spores ripe in May or June. — On shaded soil or rocks. Southern Canada and northern United States south in the East to Maryland; Europe and Siberia.

ILLUSTRATION: Grout, *Moss Flora of North America*, vol. 2, Pl. 101C.

Upper Peninsula: Chippewa and Keweenaw counties, *Steere.* Ontonagon Co., *Nichols & Steere.*
Lower Peninsula: Alpena Co., *Hall.* Cheboygan Co., *Nichols.* Emmet Co., *Steere.*

7. *Mnium cuspidatum* Hedw. — Sterile stems stoloniform; fertile stems erect; leaves obovate, acute or short-acuminate, bordered,

singly serrate in the upper part only; costa percurrent; cells up to 25 µ, irregularly rounded-hexagonal, thickened at the corners; synoicous; sporophytes single; spores ripe in late spring. — On soil or rotten logs, in woods and also in fields. British Columbia to Arizona and widespread in eastern North America; Europe and Asia.

ILLUSTRATIONS: Conard, *How to Know the Mosses* (ed. 2), fig. 105d, 154a–d. Grout, *Mosses with Hand-Lens and Microscope*, fig. 117d–f (as *M. sylvaticum*). Jennings, *Mosses of Western Pennsylvania* (ed. 2), Pl. 25. Welch, *Mosses of Indiana*, fig. 129.

Upper Peninsula: Baraga Co., *Parmelee*. Chippewa Co., *Thorpe*. Gogebic Co., *Golley*. Houghton Co., *Parmelee*. Marquette Co., *Nichols*. Ontonagon Co., *Darlington*.
Lower Peninsula: Alpena Co., *Hall*. Cheboygan Co., *Nichols*. Clinton Co., *Parmelee*. Eaton Co., *Darlington*. Ingham Co., *Wheeler*. Kalamazoo Co., *Darlington*. Lenawee Co., *Stearns*. Macomb Co., *Cooley*. Oakland and St. Joseph counties, *Darlington*. Washtenaw Co., *Kauffman*.

8. *Mnium medium* BSG — Sterile stems stoloniform; fertile stems erect; leaves broadly oval to ± obovate, short-cuspidate, bordered, singly toothed throughout (the teeth 1-celled); synoicous; sporophytes clustered; spores maturing in April and May. — On soil in wet places. Greenland; across the continent, south to Maryland in the East, California in the West; Europe and Asia.

ILLUSTRATIONS: Grout, *Moss Flora of North America*, vol. 2, Pl. 101. Jennings, *Mosses of Western Pennsylvania* (ed. 2), Pl. 25. Welch, *Mosses of Indiana*, fig. 130.

Upper Peninsula: Chippewa Co., *Steere*. Gogebic Co., *Golley*. Marquette Co., *Nichols*. Ontonagon Co., *Nichols & Steere*.
Lower Peninsula: Cheboygan Co., *Nichols*. Washtenaw Co., *Kauffman*.

9. *Mnium affine* Bland. — Plants up to 3 cm. or more high, usually brown-tomentose below; sterile stems commonly elongate and horizontal; leaves broadly oval to obovate, short-cuspidate, toothed throughout with single teeth (sometimes greatly reduced); costa ending in the cuspidate apex; cells irregularly hexagonal, usually elongate and arranged in oblique rows, up to 50 µ or more; dioicous; setae sometimes clustered; spores in April and May. — On soil in swampy places. Greenland; almost throughout North America; Europe, North Africa, Macaronesia, and Asia.

ILLUSTRATIONS: Conard, *How to Know the Mosses* (ed. 2), fig. 154c–d. Grout, *Mosses with Hand-Lens and Microscope*, Pl. 49. Jennings, *Mosses of Western Pennsylvania* (ed. 2), Pl. 26 (as var. *rugicum*). Welch, *Mosses of Indiana*, fig. 131.

Upper Peninsula: Keweenaw Co., *Allen & Stuntz*. Mackinac Co., *Kauffman*. Ontonagon Co., *Nichols & Steere*.

Lower Peninsula: Alpena Co., *Hall.* Cheboygan Co., *Nichols.* Leelanau Co., *Darlington.* Macomb Co., *Cooley.*

9a. *Mnium affine* var. *ciliare* (Grev.) C. M. — Leaves sharply toothed, the teeth 2–4 cells long. — Ranging with the species.

Upper Peninsula: Chippewa Co., *Thorpe.* Ontonagon Co., *Darlington* & *Bessey.*
Lower Peninsula: Cheboygan Co., *Nichols.* Clinton and Ingham counties, *Parmelee.* Leelanau Co., *Darlington.* Macomb Co., *Cooley.* Washtenaw Co., *Kauffman.*

10. *Mnium cinclidioides* Hüb. — Loosely tufted plants up to 5 cm. or more high; leaves large, obovate, rounded at apex or ending in a short, blunt point; margin indistinct, entire or nearly so; costa ending below the apex; cells large, about 4:1, elongate in an oblique direction; dioicous; spores in May or June. — In swamps and bogs. Greenland to Alaska and south to Pennsylvania, the Great Lakes, and Montana; Europe and Asia.

ILLUSTRATIONS: Grout, *Mosses with Hand-Lens and Microscope,* fig. 120. Jennings, *Mosses of Western Pennsylvania* (ed. 2), Pl. 27.

Upper Peninsula: Alger and Chippewa counties, *Steere.* Houghton Co., *Parmelee.* Keweenaw Co., *Povah.* Marquette Co., *Nichols.* Ontonagon Co., *Nichols* & *Steere.*
Lower Peninsula: Cheboygan Co., *Ikenberry* & *Steere.*

11. *Mnium punctatum* Hedw. — Plants in loose tufts up to 5 cm. high, or more, often tomentose below; leaves oval to obovate, rounded to emarginate at apex, distinctly bordered, entire; costa ending at or near the apex, sometimes merging with the border to form a short, blunt apiculus; cells irregularly hexagonal or somewhat elongate; dioicous; spores in winter or spring. — On soil or rocks in wet places, particularly near brooks and springs. Greenland to Alaska, south to Georgia, Colorado, and California; Europe and Asia.

ILLUSTRATIONS: Conard, *How to Know the Mosses* (ed. 2), fig. 153a–b. Grout, *Mosses with Hand-Lens and Microscope,* fig. 118-a-c. Welch, *Mosses of Indiana,* fig. 132.

Upper Peninsula: Chippewa Co., *Steere.* Ontonagon Co., *Nichols* & *Steere.*
Lower Peninsula: Cheboygan Co., *Nichols.* Emmet and Grand Traverse counties, *Steere.*

11a. *Mnium punctatum* var. *elatum* Schimp. — Plants relatively robust, with large leaves, usually sterile. — Ranging with the species.

ILLUSTRATION: Grout, *Mosses with Hand-Lens and Microscope,* fig. 118d–e.

[98]

Upper Peninsula: Alger Co., *Wheeler.* Chippewa Co., *Thorpe.* Houghton Co., *Parmelee.* Keweenaw Co., *Cooper.* Marquette Co., *Nichols.* Menominee Co., *Hill.* Ontonagon Co., *Darlington* & *Bessey.*
Lower Peninsula: Charlevoix Co., *Parmelee.* Cheboygan Co., *Wheeler.* Emmet and Grand Traverse counties, *Steere.* Ingham and Leelanau counties, *Darlington.* Muskegon Co., *Steere.* Van Buren and Washtenaw counties, *Kauffman.*

12. *Mnium pseudopunctatum* Bruch & Schimp. — Differing from *M. punctatum* in having a synoicous inflorescence and costa ending somewhat below the apex. — Greenland; across the continent, south to the Great Lakes area; Europe and Asia.

Lower Peninsula: Emmet Co. (Cecil Bay), *Sharp.*

13. *Mnium andrewsianum* Steere. — Similar to *M. punctatum;* leaves very small with the costa ending well below the apex; dioicous. — On soil in swampy places. Greenland; arctic America, south to Ontario and Michigan; Siberia.

ILLUSTRATION: Steere, Bryol. 61: 179, fig. 1–12.

Upper Peninsula: Alger Co. (near Munising), *Sharp.*

2. *Cinclidium* Sw.

Similar to *Mnium,* differing mainly in the shorter, blunt teeth of the exostome and the endostome consisting of a dome-like structure supported by segments alternating with the teeth; leaves bordered and entire.

Cinclidium stygium Sw. — Plants green or red-brown above, becoming black and tomentose below, up to 3 cm. high, or more; leaves obovate, sharply apiculate, strongly bordered, entire; costa generally extending into the apiculus; cells irregularly elongate-hexagonal; synoicous. — In swamps or bogs. Greenland; across arctic America, south to Michigan; Europe and Asia.

ILLUSTRATIONS: Figure 78. Grout, *Moss Flora of North America,* vol. 2, Pl. 106.

Upper Peninsula: Mackinac Co., *Steere.*
Lower Peninsula: Emmet and Presque Isle counties, *Nichols.*

AULACOMNIACEAE

Plants loosely or densely tufted, usually tomentose, erect, branching by yearly innovations; leaves oblong or lanceolate, rounded-obtuse to acute, not bordered; costa single, well developed; cells small, mostly unipapillose; setae elongate, terminal; capsules mostly

[99]

inclined to horizontal, sometimes asymmetric, plicate when dry; annulus broad; peristome double, complete; calyptra cucullate.

Aulacomnium Schwaegr.

Plants slender to robust, mostly yellowish-green or yellowish-brown, radiculose; leaves oblong-lanceolate to ovate or elliptic, rounded-obtuse to acute, concave or keeled; costa ending somewhat below the apex; cells small, rounded or short-elliptic, incrassate, unipapillose on both surfaces; capsules inclined to horizontal, subcylindric, somewhat curved, plicate; peristome teeth 16, papillose, endostome consisting of a high basal membrane, narrow segments, and well-developed cilia.

1. Leaves oblong-ovate, strongly toothed in the
   upper 1/2 or 2/3 . . . . 1. *A. heterostichum*
1. Leaves lanceolate, entire or ± toothed at apex 2
   2. Basal leaf cells not swollen; brood-
      bodies fusiform . . . . 3. *A. androgynum*
   2. Basal cells swollen; brood-bodies leaf-like . . 2. *A. palustre*

1. *Aulacomnium heterostichum* (Hedw.) BSG — Plants green above, yellow-brown below; leaves oblong-ovate, obtuse and apiculate to subacute, coarsely toothed in the upper half or more, not crisped when dry, concave, often somewhat homomallous; spores in late spring or summer. — On moist, shaded earth, particularly on sandy banks. Widespread in eastern North America.

ILLUSTRATIONS: Conard, *How to Know the Mosses* (ed. 2), fig. 143. Grout, *Mosses with Hand-Lens and Microscope*, fig. 100. Jennings, *Mosses of Western Pennsylvania* (ed. 2), Pl. 28. Welch, *Mosses of Indiana*, fig. 109.

Lower Peninsula: Barry Co., *Gilly* & *Parmelee*. Cheboygan Co., *Steere*. Clinton Co., *Parmelee*. Eaton and Ingham counties, *Wheeler*. Leelanau Co., *Darlington*. Lenawee Co., *Stearns*. Oakland Co., *Darlington*. Washtenaw Co. *Kauffman*.

2. *Aulacomnium palustre* (Hedw.) Schwaegr. — Plants yellow-green or yellow-brown, often bearing stalked clusters of leaf-like brood-bodies at the tip of sterile stems; leaves lanceolate, acute or acuminate, ± denticulate at apex, crisped or contorted when dry; basal cells somewhat inflated; spores in summer. — Common on moist soil, particularly in swamps and bogs and at lake margins. Nearly throughout North America; Europe, Asia, Patagonia, and the Antipodes.

ILLUSTRATIONS: Figure 79. Conard, *How to Know the Mosses* (ed. 2), fig. 85a-d. Grout, *Mosses with Hand-Lens and Microscope*. fig. 101, 102. Jennings, *Mosses of Western Pennsylvania* (ed. 2), Pl. 28. Welch, *Mosses of Indiana*, fig. 110.

Upper Peninsula: Chippewa Co., *Thorpe.* Delta Co., *Gilly & Parmelee.* Houghton Co., *Parmelee.* Keweenaw Co., *Allen & Stuntz.* Marquette Co., *Nichols.* Ontonagon Co., *Nichols & Steere.*

Lower Peninsula: Alpena Co., *Wheeler.* Barry Co., *Gilly & Parmelee.* Cheboygan Co., *Nichols.* Eaton Co., *Darlington.* Ingham Co., *Wheeler.* Leelanau Co., *Darlington.* Washtenaw Co., *Kauffman.*

3. *Aulacomnium androgynum* (Hedw.) Schwaegr. — Similar to *A. palustre* but usually smaller, with brood-bodies fusiform; basal cells not inflated; spores maturing in summer. — On moist soil and rotten wood. Newfoundland to British Columbia, south to New York, Michigan, and Idaho; Europe, Canary Islands, Japan, and Patagonia.

ILLUSTRATIONS: Conard, *How to Know the Mosses* (ed. 2), fig. 85e–g. Grout, *Moss Flora of North America*, vol. 2, Pl. 66c.

Upper Peninsula: Marquette Co., *Nichols.*
Lower Peninsula: Cheboygan Co., *Nichols.*

MEESIACEAE

Bog plants, gregarious or tufted, erect, sparsely branched, radiculose or tomentose below; leaves of varied shapes, sometimes clearly seriate, not bordered; costa single, well developed; upper cells rounded to oblong-hexagonal, smooth or mammillose; setae terminal, elongate; capsules suberect or inclined from a long, straight neck, oblong-pyriform, strongly curved and asymmetric, smooth; peristome double, the teeth sometimes shorter than the segments; calyptra cucullate.

1. Leaves strongly squarrose with recurved tips;
   cells mammillose .   .   .   .   1. *Paludella*
1. Leaves erect to ± squarrose, not recurved at tips;
   cells smooth   2
   2. Cells large, lax and thin-walled   .   .   .   .   2. *Amblyodon*
   2. Cells small and firm-walled or incrassate   .   .   .   3. *Meesia*

1. *Paludella* Brid.

Plants fairly robust, in dull, pale-green to brownish tufts, densely radiculose below; leaves in 5 distinct rows, obovate, broadly acuminate, long-decurrent, very strongly squarrose-recurved, strongly keeled; margins recurved below, serrulate above; costa ending below the apex; upper cells rounded-hexagonal, mammillose; dioicous; peristome teeth as long as the segments; cilia of endostome none or reduced to stubs. — A genus of a single species: *P. squarrosa* (Hedw.) Brid. — In bogs. Greenland; Canada, Alaska, and northern United States; Europe and Asia.

Upper Peninsula: Alger Co., *Steere.* Keweenaw Co., *Holt.*
Lower Peninsula: Montcalm Co., *Schnooberger.*

2. Amblyodon BSG

Rather small, gregarious or loosely tufted, light-green plants, radiculose only at base; leaves erect-spreading when moist, shrunken and appressed when dry, oblong-lanceolate to obovate, rather abruptly sharp-pointed, plane and entire at margins; costa ending well below the apex; cells lax and thin-walled, smooth, oblong-hexagonal; endostome longer than the exostome, without cilia. — A genus of one species: A. *dealbatus* (Hedw.) BSG — On wet soil. Greenland; across the continent and south to Nova Scotia, the Great Lakes, and Colorado; Europe and Asia.

Upper Peninsula: Mackinac Co., *Sharp.*
Lower Peninsula: Kalamazoo Co., *Becker.*

3. Meesia Hedw.

Plants rather slender to fairly robust, in loose or dense, dull tufts, densely radiculose below; leaves erect- or wide-spreading, sometimes clearly 3-ranked, oblong-lanceolate to lance-ligulate, acute to rounded-obtuse; costa broad at base, disappearing below the apex to short-excurrent; cells rectangular or oblong-hexagonal, firmwalled, smooth; exostome teeth much shorter than the endostome; cilia rudimentary.

Leaves wide-spreading or squarrose, in 3 rows, acute;
  margins plane, serrate above . . . . 1. *M. tristicha*
Leaves erect, not clearly ranked, obtuse; margins
  revolute, entire . . . . 2. *M. uliginosa*

1. *Meesia tristicha* BSG — Plants dark-green, rather robust; leaves in 3 distinct rows, wide-spreading, ovate-lanceolate, subclasping at base, decurrent, acute; margins plane, serrate above; costa subpercurrent to short-excurrent; spores in summer. — In open boggy places or wet woods. Greenland; across the continent and southward to New Jersey, Ohio and the Great Lakes in the East, California in the West; Europe and Asia.

Upper Peninsula: Alger Co., *Nichols*. Keweenaw Co., *Allen & Stuntz*. Mackinac Co., *Nichols*.
Lower Peninsula: Cheboygan Co., *Nichols*. Emmet and Washtenaw counties, *Steere*.

2. *Meesia uliginosa* Hedw. — Plants smaller; leaves erect, linear-ligulate or lanceolate-ligulate, subacute or, more often, rounded-obtuse; margins revolute, entire; costa ending below the apex; spores in summer. — In bogs and wet crevices of cliffs. Greenland; across the continent, south to New York, the Great Lakes, Colorado, and California; Europe and Asia.

ILLUSTRATION: Grout, *Moss Flora of North America*, vol. 2, Pl. 64.

Upper Peninsula: Alger, Baraga, and Keweenaw counties, *Steere*.
Lower Peninsula: Cheboygan Co., *Nichols*. Emmet Co., *Steere*. Presque Isle Co., *Brown*.

CATOSCOPIACEAE

Small plants in dense, dark, brownish-green tufts; stems erect, densely radiculose below; leaves lance-acuminate; costa single, sub-percurrent; cells smooth, subquadrate; dioicous; setae terminal, elongate; capsules very small, dark, shiny, smooth, strongly asymmetric and curved so that the mouth is directed downward; annulus none; peristome single. — A family of one genus, *Catoscopium* Brid., and one species, *C. nigritum* (Hedw.) Brid. — On wet soil, lake shores and swales. Greenland; across the continent, south to Quebec, the Great Lakes, Iowa, and Montana; Europe and Asia.

ILLUSTRATION: Figure 83. Grout, *Moss Flora of North America*, vol. 2, Pl. 67E.

Upper Peninsula: Alger Co., *Nichols*.
Lower Peninsula: Cheboygan and Emmet counties, *Steere*. Presque Isle Co., *Brown*.

BARTRAMIACEAE

Small to robust, caespitose plants, erect, variously branched, often tomentose; leaves linear to ovate-lanceolate, often sheathing at base, acute or acuminate, serrate; costa single, well developed; cells sub-quadrate to rectangular or linear, papillose because of projecting end walls or rarely smooth; setae short or elongate, terminal; capsules globose, asymmetric, furrowed when dry, mostly inclined or cernuous; operculum convex to conic; peristome double, single and reduced, or lacking.

1. Leaf cells smooth; stems ± 3-angled   .   .   .   .   1. *Plagiopus*
1. Leaf cells papillose; stems 5–many-angled   2

2. Leaves linear or subulate from a ± clasping
base; stems simple or forked . . . . . . 2. *Bartramia*
2. Leaves lanceolate to ovate, not clasping at base;
branches often whorled below the inflorescences . 3. *Philonotis*

1. *Plagiopus* Brid.

Dull, dark-green or brownish plants in dense tufts; stems forked; leaves somewhat contorted but not crisped when dry, lanceolate, acuminate; margins recurved, serrate above; costa percurrent or shortly excurrent; cells subquadrate or short-rectangular, obscurely striate but not papillose; setae elongate; capsules erect or slightly inclined; peristome double; operculum conic. — Represented in North America by a single species, *P. oederi* (Brid.) Limpr. — On walls or crevices of moist, shaded calcareous cliffs. Greenland; Labrador to Alaska, south to Pennsylvania, Colorado, and Washington; Europe, Asia and Hawaii.

ILLUSTRATIONS: Figure 84. Conard, *How to Know the Mosses* (ed. 2), fig. 111. Grout, *Mosses with Hand-Lens and Microscope*, fig. 104d, g (as *Bartramia*), *Moss Flora of North America*, vol. 2, Pl. 67G.

Upper Peninsula: Delta Co., *Nichols.* Houghton Co., *Hyypio.* Keweenaw Co., *Steere.* Mackinac Co., *Nichols.* Ontonagon Co., *Steere.*

Lower Peninsula: Alpena Co., *Robinson & Wells.* Cheboygan and Emmet counties, *Steere.*

2. *Bartramia* Hedw.

Plants in dull, green or glaucous tufts; stems forked, often tomentose; leaves linear or lanceolate from a ± clasping or strongly sheathing base; costa percurrent or excurrent; cells quadrate to oblong or linear, papillose because of projecting ends; setae short to elongate; capsules ovoid or subglobose, asymmetric, furrowed, suberect or inclined; peristome double, single, or lacking (the endostome often poorly developed).

Leaves erect-spreading, enlarged and strongly sheathing at
base; upper cells obscure and partially bistratose . 2. *B. ithyphylla*
Leaves spreading, crisped when dry, subclasping at
base; upper cells unistratose . . . . . 1. *B. pomiformis*

1. *Bartramia pomiformis* Hedw. — Plants in soft, green, yellowish, or glaucous tufts; leaves slenderly acuminate from a broader, subclasping base, flexuose, crisped when dry; margins revolute; costa excurrent; cells unistratose; peristome double; spores in spring. — On banks or cliffs, in shade. Greenland; Nova Scotia to British Columbia, south to Florida, Colorado, and Oregon; Europe, Macaronesia, and Asia.

[104]

ILLUSTRATIONS: Figure 85. Conard, *How to Know the Mosses* (ed. 2), fig. 89a–c. Grout, *Mosses with Hand-Lens and Microscope*, Pl. 44. Jennings, *Mosses of Western Pennsylvania* (ed. 2), Pl. 28. Welch, *Mosses of Indiana*, fig. 111.

Upper Peninsula: Alger Co., *Wheeler*. Baraga and Houghton counties, *Parmelee*. Keweenaw Co., *Steere*. Luce and Marquette counties, *Nichols*. Ontonagon Co., *Darlington & Bessey*.

Lower Peninsula: Alpena Co., *Robinson & Wells*. Cheboygan Co., *Nichols*. Clinton Co., *Parmelee*. Ingham Co., *Wheeler*. Jackson Co., *Marshall*. Leelanau Co., *Darlington*. Lenawee Co., *Stearns*. Van Buren and Washtenaw counties, *Kauffman*.

2. *Bartramia ithyphylla* Brid. — Leaves rigid, erect-spreading, not much altered on drying, abruptly subulate from a larger, pale, strongly sheathing base; margins plane; costa indistinct above, filling most of the upper part of the leaf; cells of upper leaf partially bistratose; peristome double; spores in summer. — On damp soil and in rock crevices. Greenland; northern North America, south to Pennsylvania, the Great Lakes, Arizona, and California; Europe, Africa and Asia.

ILLUSTRATIONS: Conard, *How to Know the Mosses* (ed. 2), fig. 89d. Grout, *Moss Flora of North America*, vol. 2, Pl. 67H.

Upper Peninsula: Keweenaw Co., *Steere*.

3. *Philonotis* Brid.

Plants in dense, pale-green or yellowish, usually tomentose tufts in wet places; stems often red, frequently producing whorls of branches below the male inflorescences; leaves lanceolate or ovate, generally acute or acuminate; margins often revolute; costa percurrent to excurrent; cells oblong to linear, papillose at 1 or both ends; setae elongate; capsules inclined or cernuous; peristome double.

1. Cells papillose only at the upper ends . . . 1. *P. marchica*
1. Cells papillose at the lower ends, or sometimes
   at both ends 2
   2. Leaves doubly serrate, with margins
      revolute . . . . 2. *P. fontana*
   Leaves singly serrate,
      plane-margined . . . . . 3. *P. caespitosa* var. *compacta*

1. *Philonotis marchica* (Hedw.) Brid. — Bright-green, rather slender plants; leaves lanceolate to ovate-lanceolate; margins plane or nearly so; costa short-excurrent; cells oblong-linear, papillose at the upper ends. — On soil in wet places. Eastern North America; Europe, Macaronesia, and Asia.

[105]

ILLUSTRATIONS: Grout, *Moss Flora of North America*, vol. 2, Pl. 69G. Welch, *Mosses of Indiana*, fig. 112.

Upper Peninsula: Alger Co., *Nichols*.
Lower Peninsula: Cheboygan Co., *Nichols*.

2. *Philonotis fontana* (Hedw.) Brid. — Bright-green, yellowish, or glaucous, relatively robust plants; leaves ± plicate, lanceolate or ovate-lanceolate; margins revolute, doubly serrate; cells oblong-linear, papillose at the lower ends, sometimes at both ends. — On wet soil, sometimes under dripping water or emergent from shallow water. Greenland; widespread in North America; Mexico; Europe, Africa, Macaronesia, and Asia.

ILLUSTRATIONS: Figure 86. Conard, *How to Know the Mosses* (ed. 2), fig. 87. Grout, *Mosses with Hand-Lens and Microscope*, fig. 103. Jennings, *Mosses of Western Pennsylvania* (ed. 2), Pl. 29. Welch, *Mosses of Indiana*, fig. 113.

Upper Peninsula: Alger Co., *Wheeler*. Chippewa Co., *Steere*. Keweenaw Co., *Hermann*. Marquette Co., *Nichols*.
Lower Peninsula: Cheboygan and Emmet counties, *Nichols*. Leelanau Co., *Darlington*. Washtenaw Co., *Kauffman*.

3. *Philonotis caespitosa* Wils. var. *compacta* Dism. — Plants fairly robust, yellowish- or glaucous-green; leaves crowded, spreading, gradually acute or acuminate from a non-plicate, ovate base; margins plane, singly serrate; costa subpercurrent to shortly excurrent; cells rectangular, papillose at the lower ends or sometimes at both ends. — Connecticut and Michigan.

ILLUSTRATION: Grout, *Moss Flora of North America*, vol. 2, Pl. 69E.

Upper Peninsula: Marquette Co. (Huron Mts.), *Nichols*.

TIMMIACEAE

Rather coarse and often robust plants (resembling *Polytrichum* or *Atrichum*); stems erect, simple or forked; leaves keeled, lanceolate from a sheathing, pale or colored base; margins plane, coarsely toothed above; costa stout, prominent at back, percurrent or nearly so; upper cells small, rounded-quadrate or hexagonal, bulging on the upper surface, smooth or papillose on the lower, often bistratose near the costa; lower cells lax, hyaline, orange, or red; setae terminal, elongate; capsules suberect to inclined, oblong-cylindric; annulus broad; peristome double, the endostome consisting of a high basal membrane and 64 papillose cilia (mostly attached in 2's and 4's). — A family of one genus, *Timmia* Hedw.

Sheathing bases of leaves hyaline or yellowish . 1. *T. megapolitana*
Sheathing bases of leaves orange . . . . . 2. *T. austriaca*

1. *Timmia megapolitana* Hedw. — Leaves crisped when dry, dark-green with hyaline or yellowish sheathing bases; costa smooth at back; spores in late spring. — On soil in moist, shaded places, particularly along streams. Greenland; across Canada and Alaska, south to New Jersey and Missouri in the East, California and Arizona in the West; Europe and Asia.

ILLUSTRATIONS: Conard, *How to Know the Mosses* (ed. 2), fig. 55. Grout, *Mosses with Hand-Lens and Microscope*, Pl. 43. Jennings, *Mosses of Western Pennsylvania* (ed. 2), Pl. 29 (as *T. cucullata*). Welch, *Mosses of Indiana*, fig. 108.

Upper Peninsula: Mackinac and Marquette counties, *Nichols*.
Lower Peninsula: Alpena Co., *Robinson & Wells.* Cheboygan Co., *Nichols.* Clinton Co., *Parmelee.* Leelanau Co., *Darlington.* Van Buren and Washtenaw counties. *Kauffman.*

2. *Timmia austriaca* Hedw. — Leaves imbricate or ± crisped when dry; sheathing bases of leaves orange; costa usually toothed at back near the tip; spores in summer. — On soil, especially in rock crevices. Greenland; across Canada and Alaska, south to New Mexico and the Great Lakes; Europe and Asia.

ILLUSTRATION: Figure 87.

Upper Peninsula: Alger Co., *Steere.* Baraga Co., *Parmelee.* Keweenaw Co., *Steere.*

ORTHOTRICHACEAE

Plants mostly small, gregarious or tufted, on bark or rock, erect or with erect branches from a creeping stem; leaves mostly ovate to lanceolate, rounded-obtuse to acute or acuminate, sometimes piliferous, often crisped when dry; margins mostly recurved; costa single, well developed; upper cells mostly rounded-quadrate with thick walls, smooth, mammillose, or papillose; basal cells oblong to linear; setae short to long, terminal or lateral; capsules erect and symmetric, often immersed, smooth or furrowed; peristome double or rarely single or lacking; calyptra cucullate or mitrate, often hairy.

1. Stems long and creeping; branches erect . . . 4. *Drummondia*
1. Stems erect 2
  2. Calyptra cucullate, naked; peristome none . . 3. *Amphidium*
  2. Calyptra mitrate, mostly hairy; peristome
    present (nearly always double) 3
3. Cells at basal margins of leaf short and pale forming
  a short border . . . . 2. *Ulota*
3. Cells at basal margins of leaf not noticeably
  differentiated . . . . 1. *Orthotrichum*

1. *Orthotrichum* Hedw.

Plants mostly small, usually in small, dark-colored cushions on rocks or trees; stems short, simple or forked; leaves crowded, mostly imbricate or sometimes ± crisped when dry, lanceolate or elliptic, rounded-obtuse to acuminate, very rarely awned, mostly keeled; margins almost always recurved; costa well-developed, generally ending somewhat below the apex; cells rounded, firm-walled or incrassate, pluripapillose; basal cells rectangular, not differentiated at the margins; capsules immersed to emergent, sometimes exserted, erect, cylindric, often 8-ribbed; peristome consisting of 8 or 16 teeth and 8 or 16 narrow segments, rarely single or lacking; calyptra mitrate, usually hairy.

1. Leaves concave, with erect margins, obtuse or
   rounded-obtuse . . . . 5. *O. obtusifolium*
1. Leaves keeled, with ± recurved margins, obtuse to
   acute (not rounded) 2
     2. Stomata of exothecium superficial 3
     2. Stomata immersed 6
3. Capsules strongly ribbed when dry and empty; peristome
   teeth conspicuously perforated above 4
3. Capsules smooth or only obscurely ribbed; peristome
   teeth not much perforated 5
     4. Capsules immersed or slightly emergent . . 3. *O. sordidum*
     4. Capsules clearly emergent, sometimes
       nearly exserted . . . . 2. *O. affine*
5. Peristome teeth 16; endostome segments
   slender, smooth . . . . 1. *O. macounii*
5. Peristome teeth 8; segments rather stout, papillose . 4. *O. elegans*
     6. Growing on rock; peristome teeth mostly
       erect to spreading when dry 7
     6. Growing on trees; peristome teeth mostly
       reflexed when dry 9
7. Capsules exserted . . . . 6. *O. anomalum*
7. Capsules immersed or emergent 8
     8. Capsules immersed, abruptly narrowed to the seta 8. *O. lescurii*
     8. Capsules emergent, tapered to the seta . . 7. *O. strangulatum*
9. Leaves acute or narrowly obtuse, some leaves
   minutely apiculate . . . . 11. *O. pumilum*
9. Leaves mostly obtuse, not apiculate 10
     10. Capsules dark-brown, strongly contracted below
       the mouth when dry and empty . . . . т. *O. stellatum*
     10. Capsules straw-colored, only slightly contracted
       below the mouth . . . . 10. *O. ohioense*

1. *Orthotrichum macounii* Aust. — Plants in dense, olive-green to brownish tufts up to 3 cm. high; leaves loosely imbricate when dry, ovate-lanceolate or lanceolate, slenderly acute; costa subpercurrent; setae 2–4 mm.; capsules exserted, smooth or slightly ribbed when old; peristome teeth 16, coarsely papillose, sometimes cross-striate

at base; segments 8, smooth or nearly so; spores in summer. — On rocks. Alaska to Utah; Newfoundland.

ILLUSTRATION: Grout, *Moss Flora of North America*, vol. 2, Pl. 46A.

Upper Peninsula: Keweenaw Co. (Isle Royale), *Conard; Povah.*

2. *Orthotrichum affine* Brid. — Plants in loose tufts up to 3 cm. high; leaves imbricate when dry, oblong-lanceolate, subacute; costa percurrent or nearly so; capsules emergent or nearly exserted, strongly ribbed; peristome teeth 16, papillose, perforate above; segments 8, papillose; spores in early summer. — On trees, rarely on rocks. Michigan; Rocky Mts. and westward; Europe, North Africa, and Kamchatka.

ILLUSTRATION: Grout, *Moss Flora of North America*, vol. 2, Pl. 47.

Upper Peninsula: Keweenaw Co., *Povah.* Marquette Co., *Nichols.*
Lower Peninsula: Cheboygan and Emmet counties, *Nichols.*

3. *Orthotrichum sordidum* Sull. & Lesq. — Very close to *O. affine*: leaves more abruptly acute; capsules immersed or slightly emergent, shorter (1.2–1.6 mm.), on a shorter seta (0.5 mm.), lighter-colored (light-yellow), less strongly furrowed; spores in spring. — On trees. Southeastern Canada and northeastern United States.

ILLUSTRATIONS: Grout, *Mosses with Hand-Lens and Microscope*, Pl. 39. Welch, *Mosses of Indiana*, fig. 99.

Upper Peninsula: Chippewa Co., *Steere.* Houghton and Keweenaw counties, *Parmelee.*
Lower Peninsula: Alpena Co., *Robinson & Wells.* Cheboygan Co., *Nichols.* Leelanau Co., *Darlington.*

4. *Orthotrichum elegans* Hook. & Grev. — Plants loosely tufted, dark-green or brown, rarely more than 1 cm. high; leaves loosely imbricate when dry, oblong-lanceolate, acute or acuminate; costa subpercurrent; capsules nearly or quite exserted, smooth or only slightly furrowed when dry and empty; peristome teeth 8, strongly papillose; segments 8, stout, coarsely papillose. (Including Michigan reports of *O. speciosum.*) — On bark, occasionally on rocks. Southeastern Canada and northeastern United States.

ILLUSTRATIONS: Conard, *How to Know the Mosses* (ed. 2), fig. 69a–b (as *O. speciosum*). Grout, *Moss Flora of North America*, vol. 2, Pl. 50D.

Upper Peninsula: Chippewa and Keweenaw counties, *Steere.* Mackinac Co., *Nichols.* Ontonagon Co., *Darlington & Bessey.*
Lower Peninsula: Cheboygan and Emmet counties, *Nichols.* Leelanau Co., *Darlington.* Oscoda Co., *Parmelee.*

5. *Orthotrichum obtusifolium* Brid. — Small plants in loose, rigid tufts, usually less than 1 cm. high; brood-bodies commonly produced on the leaves; leaves closely appressed when dry, oblong-lingulate, obtuse or rounded-obtuse, very concave with erect margins; costa ending below the apex; capsules immersed or emergent, strongly ribbed and contracted below the mouth when dry; peristome teeth 16, finely papillose; segments 8; spores in spring. — On bark. Southern Canada, south to Maryland and Arizona; Europe and Asia.

ILLUSTRATIONS: Conard, *How to Know the Mosses* (ed. 2), fig. 67a–e. Grout, *Mosses with Hand-Lens and Microscope*, fig. 91. Jennings, *Mosses of Western Pennsylvania* (ed. 2), Pl. 63. Welch, *Mosses of Indiana*, fig. 100.

Upper Peninsula: Chippewa and Keweenaw counties, *Steere*. Ontonagon Co., *Nichols & Steere*.
Lower Peninsula: Alpena Co., *Robinson & Wells*. Cheboygan Co., *Nichols*. Oscoda Co., *Parmelee*. Van Buren Co., *Kauffman*. Washtenaw Co., *Steere*.

6. *Orthotrichum anomalum* Hedw. — Dark-green or blackish plants in rather dense tufts, 1–2 cm. high; leaves imbricate when dry, oblong- or ovate-lanceolate, narrowly obtuse to abruptly acute; costa extending nearly to the apex; capsules exserted, ribbed; peristome teeth 16, faintly marked with sinuose lines; segments usually rudimentary or lacking; spores in early spring. — On rock, especially limestone. Greenland; Canada and the northern United States, south in the mountains to New Mexico; Europe, North Africa, and Asia.

ILLUSTRATIONS: Figure 88. Conard, *How to Know the Mosses* (ed. 2), fig. 71a–c. Grout, *Mosses with Hand-Lens and Microscope*, fig. 86.

Upper Peninsula: Baraga Co., *Parmelee*. Delta and Keweenaw counties, *Steere*. Mackinac Co., *Wheeler*. Ontonagon Co., *Nichols & Steere*.
Lower Peninsula: Cheboygan and Emmet counties, *Nichols*. Ionia Co., Schnooberger.

7. *Orthotrichum strangulatum* P.-Beauv. — Plants in blackish, dense tufts up to 1 cm. high; leaves oblong-lanceolate, acute or somewhat obtuse; costa subpercurrent; capsules subcylindric, tapered to the seta, emergent, ribbed; peristome teeth 16, papillose (the papillae often joined in vague, branching lines); segments rudimentary or lacking; spores in spring. — On limestone. Widespread in eastern United States; Utah.

ILLUSTRATIONS: Conard, *How to Know the Mosses* (ed. 2), fig. 71d–e. Grout, *Moss Flora of North America*, vol. 2, Pl. 53B. Jennings, *Mosses of Western Pennsylvania* (ed. 2), Pl. 18. Welch, *Mosses of Indiana*, fig. 101.

Upper Peninsula: Alger Co., *Wheeler*. Keweenaw Co., *Steere*.
Lower Peninsula: Alpena Co., *Robinson & Wells*. Charlevoix Co., *Wheeler*.

8. *Orthotrichum lescurii* Aust. — Similar to *O. strangulatum*: leaves more slenderly lanceolate; capsules immersed, rather abruptly narrowed to the seta (globose-ovoid when moist); peristome teeth more finely papillose. — On rock. New England to British Columbia, south to Pennsylvania and Missouri.

ILLUSTRATIONS: Grout, *Moss Flora of North America,* vol. 2, Pl. 53C. Jennings, *Mosses of Western Pennsylvania* (ed. 2), Pl. 19.

Lower Peninsula: Washtenaw Co., *Lowry & Steere.*

9. *Orthotrichum stellatum* Brid. — Plants in small, dense tufts up to 1 cm. high; leaves imbricate when dry, broadly lanceolate, subacute to obtuse; costa ending just below the apex; capsules immersed to emergent, strongly ribbed, strongly contracted below the mouth when dry and empty; peristome double, the teeth 16, finely papillose; spores in late spring. — On trees. Widespread in eastern North America; Europe.

ILLUSTRATIONS: Grout, *Mosses with Hand-Lens and Microscope,* fig. 88 (as *O. strangulatum*). Jennings, *Mosses of Western Pennsylvania* (ed. 2), Pl. 19. Welch, *Mosses of Indiana,* fig. 102.

Upper Peninsula: Chippewa Co., *Steere.*
Lower Peninsula: Alpena Co., *Robinson & Wells.* Clinton Co., *Parmelee.*

10. *Orthotrichum ohioense* Sull. & Lesq. — Small plants in small, dense cushions about 5–10 mm. high; leaves imbricate when dry, oblong-lanceolate, subacute to narrowly obtuse; costa ending somewhat below the apex; capsules immersed, straw-colored, somewhat ribbed, not or slightly contracted below the mouth when dry and empty; peristome double, teeth 16, finely papillose; spores in early spring. — On trees. Widespread in eastern North America.

ILLUSTRATIONS: Conard, *How to Know the Mosses* (ed. 2), fig. 73f–g. Grout, *Mosses with Hand-Lens and Microscope,* fig. 87. Jennings, *Mosses of Western Pennsylvania* (ed. 2), Pl. 19. Welch, *Mosses of Indiana,* fig. 103.

Upper Peninsula: Chippewa Co., *Steere.*
Lower Peninsula: Alpena Co., *Robinson & Wells.* Ingham Co., *Parmelee.* Ottawa Co., *Kauffman.* Washtenaw Co., *Steere.*

11. *Orthotrichum pumilum* Dicks. — Small plants (less than 1 cm. high), in small, dense, dark-green tufts; leaves broadly or narrowly oblong-lanceolate, acute to narrowly obtuse, often ending abruptly in a single subhyaline cell; capsules immersed to slightly emergent, light-colored, ribbed, contracted below the mouth when dry; peristome teeth 16, finely papillose, segments 8; spores in spring. — On bark. Widespread in eastern North America; Utah and Idaho; Europe, North Africa, and Japan.

ILLUSTRATIONS: Conard, *How to Know the Mosses* (ed. 2), fig. 73a–e. Grout, *Mosses with Hand-Lens and Microscope*, fig. 89 (as *O. schimperi*). Welch, *Mosses of Indiana*, fig. 104.

Lower Peninsula: Cheboygan Co., *Nichols*. Ingham Co., *Parmelee*. Van Buren Co., *Kauffman*. Washtenaw Co., *Steere*.

2. *Ulota* Mohr

Small, green, yellow, or brownish plants in small tufts, radiculose below, erect, forked; leaves crowded, mostly crisped and contorted when dry, narrowly lanceolate from a broader, concave base, acute or rarely blunt; costa subpercurrent; upper cells small, rounded or elliptic, very thick-walled, papillose, those of the base linear, yellowish, incrassate, short and pale in 1–2 marginal rows; perichaetial leaves not much differentiated; setae terminal, straight, ± elongate; capsules exserted, erect and symmetric, cylindric with a long, tapering neck, 8-ribbed or merely puckered at the mouth; operculum rostrate; stomata superficial; peristome double, the teeth 16, segments 8 or 16; calyptra mitrate, plicate, hairy.

1. Leaves not or slightly contorted when dry; plants growing
   on rock . . . . 2. *U. hutchinsiae*
1. Leaves crisped and contorted when dry; plants
   growing on bark 2
   2. Setae 1–2 mm. long; capsules 8-ribbed when dry . 1. *U. crispa*
   2. Setae 3–6 mm. long; capsules puckered at the
      mouth when dry . . . . 3. *U. ludwigii*

1. *Ulota crispa* (Hedw.) Brid. — Plants in small, yellow-green tufts 5–10 mm. high; leaves lanceolate from a broader, concave base, acute or somewhat obtuse, strongly crisped when dry; costa vanishing in the apex; upper cells nearly circular or transversely elliptic, incrassate; capsules exserted on a seta up to 2 mm. long, subcylindric and 8-ribbed when dry, tapered to a long neck merging with the seta; peristome teeth 16, finely papillose; segments 8; spores in late spring or early summer. — On bark of hardwood trees. Alaska; eastern North America; Europe, Canary Islands, Asia, and Tasmania.

ILLUSTRATIONS: Conard, *How to Know the Mosses* (ed. 2), fig. 65c–e. Grout, *Mosses with Hand-Lens and Microscope*, Pl. 38. Jennings, *Mosses of Western Pennsylvania* (ed. 2), Pl. 20. Welch, *Mosses of Indiana*. fig. 106.

Upper Peninsula: Chippewa Co., *Steere*. Gogebic Co., *Bessey*. Houghton Co., *Parmelee*. Keweenaw Co., *Hermann*. Marquette Co., *Nichols*. Ontonagon Co., *Nichols* & *Steere*.

Lower Peninsula: Alpena Co., *Robinson* & *Wells*. Cheboygan Co., *Nichols*. Clinton Co., *Parmelee*. Emmet Co., *Wheeler*.

2. *Ulota hutchinsiae* (Sm.) Hamm. — Plants dark, brown or black-ish below, in rigid tufts 1–2 cm. high; leaves imbricate when dry, lanceolate or oblong-lanceolate, obtuse or subacute; costa subpercurrent; cells rounded or elliptic, incrassate; setae 2–4 mm. long; capsules exserted, subcylindric and ribbed when dry; peristome teeth 16, finely papillose; segments 8; spores in early summer. — On rocks, rarely on trees. Widespread in eastern North America; Arizona; Europe and Japan.

ILLUSTRATIONS: Figure 89. Conard, *How to Know the Mosses* (ed. 2), fig. 65a–b (as *U. americana*). Grout, *Moss Flora of North America*, vol. 2, Pl. 56C. Jennings, *Mosses of Western Pennsylvania* (ed. 2), Pl. 19 (as *U. americana*).

Upper Peninsula: Chippewa Co., *Steere.* Keweenaw Co., *Allen & Stuntz.* Marquette Co., *Nichols.*
Lower Peninsula: Emmet Co., *Steere.*

3. *Ulota ludwigii* (Brid.) Brid. — Plants in small, green or brown-ish tufts 5–10 mm. high; leaves somewhat contorted or crisped when dry, lanceolate from a broader, concave base, acute or somewhat obtuse; costa vanishing near the apex; cells subcircular to elliptic, incrassate; setae 3–6 mm. long; capsules exserted, pyriform, puckered at the mouth especially when dry and empty; peristome teeth 16, finely papillose; segments rudimentary or lacking; spores in autumn. — On bark of deciduous trees. Eastern North America; Europe.

ILLUSTRATIONS: Conard, *How to Know the Mosses* (ed. 2), fig. 65f. Grout, *Mosses with Hand-Lens and Microscope*, fig. 84. Jennings, *Mosses of Western Pennsylvania* (ed. 2), Pl. 19.

Upper Peninsula: Chippewa Co., *Steere.* Houghton Co., *Parmelee.* Mackinac and Marquette counties, *Steere.*
Lower Peninsula: Cheboygan and Emmet counties, *Nichols.*

3. *Amphidium* Schimp.

Plants in dense, dark- or yellow-green tufts on rock, radiculose; leaves crisped when dry, narrowly lanceolate; costa percurrent or nearly so; upper cells rounded-quadrate or hexagonal, incrassate; basal cells pale, rectangular; perichaetial leaves differentiated, sometimes strongly sheathing; capsules emergent to exserted, pyriform, strongly ribbed and often contracted below the mouth when dry; peristome none; calyptra cucullate, naked.

Amphidium lapponicum (Hedw.) Schimp. — Plants dark-green above, brown or blackish below, 1–3 cm. high; leaves crisped when dry, narrowly lanceolate, sharply or bluntly acute; margins plane and entire; costa ending shortly below the apex; upper cells densely papillose; autoicous; perichaetial leaves greatly enlarged; capsules about ½-emergent; spores in spring or early summer. — On boulders and cliffs. Greenland; southern Canada to California; northeastern United States; Europe and Asia.

ILLUSTRATIONS: Figure 90. Grout, *Mosses with Hand-Lens and Microscope*, fig. 82, *Moss Flora of North America*, vol. 2, Pl. 59B.

Upper Peninsula: Keweenaw and Ontonagon counties, *Steere*.

4. *Drummondia* Hook.

Plants in low, dense, dark-green mats on bark; primary stems long and creeping, producing numerous, erect, short, densely foliate branches; leaves appressed when dry, oblong- to ovate-lanceolate, acute or somewhat obtuse, keeled; margins plane or inflexed, entire; costa subpercurrent; upper cells small, rounded, incrassate, basal cells scarcely differentiated except at the insertion where they are somewhat enlarged; dioicous; setae 2–3 mm. long; capsules exserted, ovoid, smooth; peristome single, the teeth short, truncate, smooth; calyptra cucullate, naked. — A small genus represented by a single species in North America, *D. prorepens* (Hedw.) E. Britt. — On bark. Widespread in eastern North America.

ILLUSTRATIONS: Figure 91. Conard, *How to Know the Mosses* (ed. 2), fig. 98a, c–d. Grout, *Mosses with Hand-Lens and Microscope*, fig. 83 (as *D. clavellata*). Jennings, *Mosses of Western Pennsylvania* (ed. 2), Pl. 18. Welch, *Mosses of Indiana*, fig. 107.

Upper Peninsula: Ontonagon Co., *Nichols & Steere*.
Lower Peninsula: Cheboygan Co., *Steere*. Ingham Co., *Wheeler*.

FONTINALACEAE

Slender to robust, ± elongate plants, submerged (at least at high water), attached at base, usually denuded below, freely branched; leaves usually in 3 rows, rarely falcate, subulate to lanceolate or suborbicular, obtuse to acuminate, often keeled; costa none or single; cells smooth, rhomboidal to elongate and prosenchymatous, the alar cells sometimes clearly differentiated; perichaetium differentiated, sometimes long-sheathing; capsules erect, immersed or shortly exserted; annulus none; operculum conic or rostrate; peristome double, the exostome of 16 teeth, the endostome of 16 cilia generally ± united by transverse strands.

Leaves costate   . . . . 1. *Fontinalis*
Leaves ecostate   . . . . 2. *Dichelyma*

[114]

1. *Fontinalis* Hedw.

Rather slender to robust, aquatic mosses with elongate, freely branched stems attached at base; leaves subulate to narrowly lanceolate or suborbicular, inserted in 3 rows, sometimes clearly seriate, ± keeled or keeled-conduplicate; costa none; cells smooth, prosenchymatous; capsules erect and symmetric, immersed or emergent; operculum conic; peristome teeth 16; endostome consisting of 16 cilia united by transverse strands to form a trellis; calyptra conic. — Growing submerged in running water.

1. Older stem leaves keeled or keeled-conduplicate, the
   younger stem and branch leaves keeled or merely concave . . . 2
1. Neither stem nor branch leaves keeled-conduplicate 5
   2. Leaf keels ± straight above the basal
      curve . . . . 1b. *F. antipyretica* var. *patula*
   2. Leaf keels ± curved above the basal curve 3
3. Leaves ovate to suborbicular, 1–2: 1, rather broadly
   obtuse; plants robust . . . . 1a. *F. antipyretica* var. *gigantea*
3. Leaves oblong to ovate-lanceolate, 1.53:1, acute to
   narrowly obtuse; plants relatively slender 4
   4. Stem leaves 5–8 mm. long, oval or ovate-lanceolate,
      usually obtuse; perichaetial leaves oval or
      suborbicular, obtuse . . . . 1. *F. antipyretica*
   4. Stem leaves 2.5–5 mm. long, oblong-lanceolate, usually
      acute; perichaetial leaves oval-lanceolate,
      apiculate . . . . 2. *F. neo-mexicana*
5. Stem and branch leaves unlike, not completely intergrading
   in size or shape 6
5. Stem and branch leaves intergrading 7
   6. Stems flaccid; leaves indistinctly dimorphous, the
      stem leaves soft, narrowly to broadly
      lanceolate . . . . 3. *F. missourica*
   6. Stems ± rigid; leaves distinctly dimorphous,
      the stem leaves rather firm, narrowly
      lanceolate . . . . 4. *F. disticha*
7. Stem leaves concave or channeled, rather firm 8
7. Stem leaves slightly concave or nearly plane, mostly soft . . . 10
   8. Stem leaves linear-lanceolate, up to 2 mm. apart;
      plants very slender and delicate . . . . 9. *F. filiformis*
   8. Stem leaves ovate to oblong-lanceolate or narrowly
      lanceolate, only slightly distant; plants not filiform 9
9. Leaves oblong or ovate-lanceolate, with margins mostly
   narrowly involute at apex . . . . . 5. *F. novae-angliae*
9. Leaves sublinear or oblong-lanceolate, with margins mostly
   broadly involute . . . . 5a. *F. novae-angliae* var. *cymbifolia*
   10. Leaves up to 8 mm. long, narrowly lanceolate;
       auricles distinct . . . . 8. *F. flaccida*
   10. Leaves up to 5.5 mm. long, narrowly lanceolate
       to ovate-lanceolate; alar cells not or
       somewhat differentiated 11
11. Leaves narrowly lanceolate to ovate-lanceolate, with apices
    mostly long and narrow; alar cells rather indistinct . 6. *F. hypnoides*
11. Leaves oblong-lanceolate to broadly ovate-lanceolate, with
    apices mostly short and broad; alar cells ± enlarged . 7. *F. duriaei*

1. *Fontinalis antipyretica* Hedw. — Rather slender, dark-green plants; leaves somewhat distant, keeled-conduplicate, oval- or ovate-lanceolate, about 2–3:1; cells oblong-hexagonal to linear-rhomboidal; perichaetial leaves oval or suborbicular, obtuse; capsules oval or oblong, 2–3 mm. long, ± emergent. — Greenland; across the continent, south to Pennsylvania and New Mexico; Europe, Asia, and South America.

ILLUSTRATIONS: Figure 92. Conard, *How to Know the Mosses* (ed. 2), fig. 251. Grout, *Moss Flora of North America*, vol. 3, Pl. 74 (fig. 14–17). Jennings, *Mosses of Western Pennsylvania* (ed. 2), Pl. 64.

Upper Peninsula: Delta Co., *Gleason, Jr.* Marquette Co., *Dodge.* Ontonagon Co., *Nichols* & *Steere.*
Lower Peninsula: Berrien Co., *Dodge.*

1a. *Fontinalis antipyretica* var. *gigantea* (Sull.) Sull. — Robust plants with broader, usually longer, and more crowded leaves. — Southern Canada and northern United States; Europe and North Africa.

ILLUSTRATIONS: Grout, *Moss Flora of North America*, vol. 3, Pl. 74 (fig. 1-3), *Mosses with Hand-Lens and Microscope*, Pl. 88d. Jennings, *Mosses of Western Pennsylvania* (ed. 2), Pl. 64.

Upper Peninsula: Alger Co., *Gillman.* Chippewa Co., *Thorpe.* Gogebic Co., *Ikenberry.* Houghton Co., *Steere.* Marquette Co., *Nichols.* Ontonagon Co., *Paul.*

1b. *Fontinalis antipyretica* var. *patula* (Card.) Welch — Differing from the species in having leaf keels straight above the basal curve; leaves plane or somewhat concave, broadly short-acuminate. — Western North America; Michigan, Nova Scotia, and Connecticut.

ILLUSTRATION: Grout, *Moss Flora of North America*, vol. 3, Pl. 74 (fig. 11–13).

Upper Peninsula: Keweenaw Co., *Steere.* Marquette Co., *Nichols.*

2. *Fontinalis neo-mexicana* Sull. & Lesq. — Plants slender, dull, green or yellowish, or darker; leaves keeled-conduplicate, the keel slightly curved and often split, ovate-lanceolate, acute; perichaetial leaves ovate-lanceolate, abruptly apiculate. — Alaska to New Mexico; Michigan; Argentina.

ILLUSTRATION: Grout, *Moss Flora of North America*, vol. 3, Pl. 73B.

Lower Peninsula: Manistee Co. (Bear Lake), *Hill.*

3. *Fontinalis missourica* Card. — Plants soft and slender, 15–20 cm. long; leaves rather soft, concave throughout or sometimes plane

above, lanceolate or narrowly ovate-lanceolate, narrowly to broadly long-acuminate. — Eastern North America.

ILLUSTRATION: Grout, *Moss Flora of North America*, vol. 3, Pl. 75 (fig. 15–19).

Upper Peninsula: Luce Co., *Nichols*. Ontonagon Co., *Nichols & Steere*.

4. *Fontinalis disticha* Hook. & Wils. — Slender and rather delicate, yellow or dull-green plants, 15–20 cm. long; stem leaves rather distant, ± firm, moderately concave, narrowly lanceolate, narrowly acuminate. —Eastern North America.

ILLUSTRATIONS: Grout, *Moss Flora of North America*, vol. 3, Pl. 75 (fig. 10–14). Welch, *Mosses of Indiana*, fig. 246.

Lower Peninsula: Cheboygan Co. (Pigeon River), *Ehlers*.

5. *Fontinalis novae-angliae* Sull. — Plants slender or somewhat robust, green or yellow-green; leaves 2.5–4 mm. long, ovate-lanceolate, ± obtuse or truncate, frequently serrulate at apex, the upper margins usually narrowly involute; cells 8–15:1; capsules immersed. — Widespread in eastern North America.

ILLUSTRATIONS: Grout, *Moss Flora of North America*, vol. 3, Pl. 76 (fig. 20–23), *Mosses with Hand-Lens and Microscope*, fig. 217 and Pl. 88b–c. Jennings, *Mosses of Western Pennsylvania* (ed. 2), Pl. 34. Welch, *Mosses of Indiana*, fig. 248.

Upper Peninsula: Baraga Co., *Michigan Dept. of Conservation*. Chippewa Co., *Steere*. Keweenaw Co., *Welch*. Mackinac Co., *Ehlers*. Marquette Co., *Nichols*. Ontonagon Co., *Steere*.
Lower Peninsula: Allegan Co., *Michigan Dept. of Conservation*. Cheboygan Co., *Nichols*. Kalamazoo Co., *Becker*.

5a. *Fontinalis novae-angliae* var. *cymbifolia* (Aust.) Welch — Many leaves deeply concave or ± canaliculate with margins often broadly involute above or nearly to the base, thus appearing sublinear. — Ranging with the species.

ILLUSTRATION: Grout, *Moss Flora of North America*, vol. 3, Pl. 76 (fig. 26–28).

Upper Peninsula: Baraga Co., *Michigan Dept. of Conservation*.

6. *Fontinalis hypnoides* C. Hartm. — Plants rather delicate, soft, bright- or yellowish-green; leaves 3.5–5 mm. long, plane or nearly so, narrowly lanceolate to ovate-lanceolate, slenderly long-acuminate, often decurrent, sometimes slightly auriculate; alar cells usually ± differentiated. — Canada and northern United States; Europe and Asia.

ILLUSTRATION: Grout, *Moss Flora of North America*, vol. 3, Pl. 77 (fig. 1–5).

Lower Peninsula: Jackson Co. (Clarke Lake), *Purpus.*

7. **Fontinalis duriaei** Schimp. — Plants soft, green to yellowish; leaves 3–5 mm. long, plane or nearly so, broadly ovate to oval-lanceolate, gradually tapered from near the middle to a broad, short, acute point, usually not decurrent, not or only slightly auriculate; alar cells ± enlarged; capsules immersed, not contracted below the mouth when dry. (Young or small plants have been called *F. nitida*, which Welch now considers a synonym.) — Widespread in North America; Europe, Asia, and Brazil.

ILLUSTRATIONS: Conard, *How to Know the Mosses* (ed. 2), fig. 253. Grout, *Moss Flora of North America*, vol. 3, Pl. 77 (fig. 12–15). Welch, *Mosses of Indiana*, fig. 250.

Upper Peninsula: Chippewa Co., *Ehlers.* Gogebic Co., *Conard.* Keweenaw Co., *Hermann.* Luce Co., *Nichols.* Mackinac Co., *Ehlers.* Marquette Co., *Nichols.* Ontonagon Co., *Steere.*
Lower Peninsula: Calhoun Co., *Tarzwell.* Cheboygan and Emmet counties, *Ehlers.* Gratiot Co., *Schnooberger.* Leelanau Co., *Darlington.* Ogemaw Co., *Wynne.*

8. **Fontinalis flaccida** Ren. & Card. — Soft and delicate, yellowish plants; most leaves plane throughout, 4–7 mm. long, narrowly ovate-lanceolate or lanceolate, narrowly to broadly acuminate, usually abruptly obtuse or truncate at apex; alar cells much enlarged in conspicuous auricles; capsules immersed, subcylindric. — Eastern United States and Canada.

ILLUSTRATION: Grout, *Moss Flora of North America*, vol. 3, Pl. 78a–d.

Upper Peninsula: Houghton Co. (Bob Lake), *Michigan Dept. of Conservation.*

9. **Fontinalis filiformis** Sull. & Lesq. — Very slender and delicate, yellowish plants with ± rigid stems; leaves 3–6 mm. long, firm when moist, deeply concave to subtubulose, very narrowly lanceolate, subulate-acuminate. — Eastern North America.

ILLUSTRATION: Grout, *Moss Flora of North America*, vol. 3, Pl. 79 (fig. 12–14).

Lower Peninsula: Gratiot Co., *Schnooberger.*

2. Dichelyma Myr.

Rather slender, branched plants in loose, stringy tufts or mats; leaves keeled, lanceolate or lance-subulate, ± falcate-secund; costa subpercurrent to long-excurrent; cells linear; perichaetial leaves long-sheathing; capsules oval to subcylindric, exserted or emerging from the side of the perichaetium; cilia of endostome free or united by transverse strands at least at the apex. — On sticks, stones, or bases of trees or shrubs in places periodically inundated.

1. Costa long-excurrent . . . . 1. *D. capillaceum*
1. Costa subpercurrent to short-excurrent 2
    2. Costa percurrent or short-excurrent; leaves subulate
    at the tips; trellis of endostome perfect . . . 2. *D. falcatum*
    2. Costa percurrent or nearly so; leaves acuminate to
    an acute or obtuse point; cilia of endostome free
    except for a few connections at apex . . . 3. *D. pallescens*

1. *Dichelyma capillaceum* (With.) Myr. — Leaves erect-spreading or somewhat falcate-secund, 5–7 mm. long, lanceolate, long-subulate; costa long-excurrent; setae 3–5 mm. long; capsule emerging from the side of the perichaetium at maturity; segments of endostome united by cross bars at apex. — Eastern North America; Europe.

ILLUSTRATIONS: Conard, *How to Know the Mosses* (ed. 2), fig. 200. Grout, *Moss Flora of North America*, vol. 3, Pl. 79 (fig. 38–42), *Mosses with Hand-Lens and Microscope*, fig. 219. Welch, *Mosses of Indiana*, fig. 251.

Upper Peninsula: Marquette Co., *Nichols*.
Lower Peninsula: Cheboygan Co., *Phinney*. Gratiot Co., *Schnooberger*. Roscommon Co., *Smith*. Van Buren Co., *Schnooberger*. Washtenaw Co., *Steere*.

2. *Dichelyma falcatum* (Hedw.) Myr. — Leaves falcate-secund, 3–5 mm. long, subulate; costa percurrent or, more often, short-excurrent; setae 5–15 mm. long; capsules emergent at the end of the perichaetium or exserted; endostome trellis perfect. — Canada, Alaska, and northern United States; Europe and Asia.

ILLUSTRATION: FIGURE 93. Grout, *Moss Flora of North America*, vol. 3, Pl. 79 (fig. 31–34).

Upper Peninsula: Keweenaw Co., *Holt*. Marquette Co., *Nichols*. Ontonagon Co., *Nichols & Steere*.

3. *Dichelyma pallescens* BSG — Leaves secund, slightly falcate, gradually acuminate, acute to obtuse at the tip; costa percurrent or nearly so; cilia of endostome free or united by cross bars at apex. — Southeastern Canada and northeastern United States.

ILLUSTRATIONS: Grout, *Moss Flora of North America*, vol. 3, Pl. 79 (fig. 43–47), *Mosses with Hand-Lens and Microscope*, fig. 220. Jennings, *Mosses of Western Pennsylvania* (ed. 2), Pl. 35.

Upper Peninsula: Alger Co., *Steere.* Keweenaw and Marquette counties, *Nichols.* Ontonagon Co., *Nichols & Steere.*
Lower Peninsula: Cheboygan Co., *Steere.* Roscommon Co., *Schnooberger.*

CLIMACIACEAE

Plants mostly dendroid, growing from an underground creeping stem; stems and branches paraphyllose; leaves strongly unicostate, sometimes decurrent; cells rhomboidal, smooth; setae elongate; capsules erect and symmetric or curved and cernuous; peristome double, well developed; cilia rudimentary or lacking.

Climacium Web. & Mohr

Plants erect and dendroid from an underground creeping stem (rarely prostrate and submerged, usually on wet soil); stems and branches bearing paraphyllia; leaves oblong-lanceolate to ovate; costa single, well developed; cells rhomboidal, smooth; setae elongate, often clustered; capsules erect, cylindric; annulus none; operculum conic-rostrate; peristome double, the teeth long, narrowly lanceolate, minutely papillose, the segments as long as the teeth, keeled and slit, arising from a low basal membrane; cilia lacking or rudimentary.

1. Plants prostrate and irregularly branched or densely
    tufted and not obviously dendroid   .   .   .   .   3. *C. kindbergii*
1. Plants distinctly dendroid   2
    2. Leaves auricled at the base; cells not more
        than 7:1; capsules 5–6:1 .   .   .   .   .   .   1. *C. americanum*
    2. Leaves not or only slightly auricled; cells 7-10:1;
        capsules 3–4:1   .   .   .   .   2. *C. dendroides*

1. *Climacium americanum* Brid. — Plants dendroid, gregarious, loosely tufted; leaves subacute, serrate above, auriculate, bisulcate; costa ending near the apex; cells 5–7:1, capsules about 6 mm. long; spores in autumn. — On wet soil and rotten logs in swamps. Widespread in eastern North America.

ILLUSTRATIONS: Conard, *How to Know the Mosses* (ed. 2), fig. 195a–c. Grout, *Mosses with Hand-Lens and Microscope*, fig. 158c. Jennings, *Mosses of Western Pennsylvania* (ed. 2), Pl. 35. Welch, *Mosses of Indiana*, fig. 133.

Upper Peninsula: Chippewa Co., Thorpe. Houghton and Keweenaw counties, *Parmelee.* Marquette Co., *Nichols.* Ontonagon Co., *Darlington.*
Lower Peninsula: Cheboygan Co., *Nichols.* Clinton Co., *Parmelee.* Eaton, Leelanau, Oakland, and St. Joseph counties, *Darlington.* Washtenaw Co., *Kauffman.*

2. *Climacium dendroides* (Hedw.) Web. & Mohr — Very similar to *C. americanum;* leaves only slightly auricled; cells longer (up to

[120]

10:1); capsules shorter (3–4:1); spores in autumn. — On wet soil in swamps, especially at the margins of lakes and streams; across the continent, south to New Jersey, New Mexico, and California; Europe and Asia.

ILLUSTRATIONS: Conard, *How to Know the Mosses* (ed. 2), fig. 195f. Grout, *Mosses with Hand-Lens and Microscope*, fig 159c–d. Welch, *Mosses of Indiana*, fig. 134.

Upper Peninsula: Ontonagon Co., *Conard*.
Lower Peninsula: Cheboygan Co., *Nichols*. Washtenaw Co., *Steere*.

3. *Climacium kindbergii* (Ren. & Card.) Grout — Plants very dark-green, sometimes nearly black, prostrate and very irregularly branched or crowded and not obviously dendroid; cells short (2–3:1), oblong-hexagonal. Probably only a form of one or both of the preceding. — In very wet places in swamps, sometimes ± submerged. Widespread in eastern United States.

ILLUSTRATIONS: Figure 94. Conard, *How to Know the Mosses* (ed. 2), fig. 195d–e. Grout, *Mosses with Hand-Lens and Microscope*, fig. 158d–e, 159a.

Upper Peninsula: Chippewa Co., *Steere*. Keweenaw Co., *Allen & Stuntz*. Ontonagon Co., *Nichols & Steere*.
Lower Peninsula: Alpena Co., *Hall*. Cheboygan Co., *Nichols*. Leelanau Co., *Darlington*. Van Buren Co., *Kauffman*. Washtenaw Co., *Steere*.

HEDWIGIACEAE

Plants loosely tufted, gray-green, prostrate, freely branched, on rock; leaves ovate, acuminate, often hyaline-tipped, concave, ecostate; margins revolute, papillose-denticulate to ciliate at the apex; costa lacking; upper cells oblong-rhombic, incrassate, papillose; lower cells elongate at the middle, quadrate toward the margins; autoicous; perichaetial leaves enlarged, ciliate at the apex; capsules immersed, subglobose, erect; operculum convex; peristome none; calyptra only covering the lid, subcucullate. — Represented in our area by a single genus, *Hedwigia* P.-Beauv., of a single species, *H. ciliata* (Hedw.) P.-Beauv. — On dry, usually exposed siliceous rocks. Widely distributed in North America; nearly cosmopolitan.

ILLUSTRATIONS: Figure 95. Conard, *How to Know the Mosses* (ed. 2), fig. 63. Grout, *Mosses with Hand-Lens and Microscope*, fig. 50, 51. Jennings, *Mosses of Western Pennsylvania* (ed. 2), Pl. 31. Welch, *Mosses of Indiana*, fig. 89.

Upper Peninsula: Baraga Co., *Parmelee*. Chippewa Co., *Steere*. Houghton and Keweenaw counties, *Parmelee*. Marquette Co., *Nichols*. Ontonagon Co., *Darlington*.

Lower Peninsula: Alpena Co., *Robinson* & *Wells*. Barry Co., *Gilly* & *Parmelee*. Cheboygan Co., *Nichols*. Clinton Co., *Parmelee*. Dickinson Co., *Wheeler*. Genesee and Ingham counties, *Darlington*. Jackson Co., *Marshall*. Macomb Co., *Gilly* & *Parmelee*. Oakland Co., *Darlington*. Washtenaw Co., *Kauffman*.

CRYPHAEACEAE

Primary stems creeping; secondary stems erect or ascending, often branched, sometimes paraphyllose; leaves imbricate when dry, ovate-lanceolate, concave, usually not plicate; costa usually single, usually well developed; cells oval or linear-elliptic, incrassate, subquadrate in numerous rows at the basal angles; sporophytes lateral; setae usually very short; capsules immersed to exserted, erect and symmetric; peristome usually double, the teeth 16, well developed; segments linear to lanceolate from a very low basal membrane (rarely lacking); calyptra often hairy.

Forsstroemia Lindb.

Secondary stems erect, sparsely to pinnately branched (resembling *Leucodon*); leaves imbricate when dry, erect-spreading when moist, oblong-ovate, short-acuminate; costa variable, lacking, short and double, or single and well developed; cells smooth, oblong-rhomboidal, rounded-quadrate in numerous rows at the basal angles; autoicous; capsules immersed to exserted, oblong-ovoid; peristome teeth linear-lanceolate; endostome rudimentary or lacking; calyptra cucullate, ± hairy.

Forsstroemia trichomitria (Hedw.) Lindb. — Green or yellow-green, loosely tufted plants, subpinnately branched; leaves somewhat plicate when dry, concave with reflexed margins, acute to abruptly short-acuminate; costa variable, short and double or lacking, sometimes single and reaching the leaf middle; perichaetial leaves long and sheathing, as long as the seta or somewhat longer; spores in late autumn to winter. — On trees and rocks. New England and Ontario, south to the Gulf of Mexico; eastern Asia.

ILLUSTRATIONS: Figure 96. Conard, *How to Know the Mosses* (ed. 2), fig. 214 (as *Leptodon*).

Lower Peninsula: Cheboygan Co. ("Mackinaw"), *Wheeler*. Ingham Co., *Beal*.

LEUCODONTACEAE

Plants mostly rather robust or coarse; primary stems creeping; secondary stems erect, horizontal, or drooping, mostly terete; leaves crowded, appressed, often plicate, ovate or ovate-lanceolate, acute or acuminate, mostly entire; costa single, double, or lacking; cells elliptic, thick-walled, smooth or sometimes projecting at the ends, subquadrate toward the basal margins; dioicous; perichaetial leaves sheathing; setae short or elongate; capsules immersed to exserted, erect and symmetric, ovoid to cylindric; peristome double or the endostome often rudimentary or lacking; calyptra cucullate, naked or hairy.

Leucodon Schwaegr.

Rather coarse and rigid plants; secondary stems julaceous, simple or sparsely branched, often curved when dry; costa lacking; capsules emergent or exserted; peristome apparently single (the endostome very rudimentary); calyptra naked.

Leucodon sciuroides (Hedw.) Schwaegr. — Plants brownish- or dull-green; secondary stems mostly simple, curved when dry, bearing clusters of minute branches in leaf axils; leaves plicate, acuminate; spores in spring. — On tree trunks, occasionally on rocks. Southeastern Canada and northeastern United States; Europe, North Africa, and Asia.

ILLUSTRATIONS: Figure 97. Grout, *Mosses with Hand-Lens and Microscope,* fig. 212g–i, *Moss Flora of North America,* vol. 3, Pl. 66.

Upper Peninsula: Alger Co., *Wheeler.* Chippewa Co., *Steere.* Houghton Co., *Parmelee.* Keweenaw Co., *Cooper.* Marquette Co., *Nichols.* Ontonagon Co., *Darlington.*

Lower Peninsula: Alpena Co., *Robinson & Wells.* Cheboygan Co., *Nichols.* Clinton Co., *Parmelee.* Emmet Co., *Gilly & Parmelee.* Leelanau Co., *Darlington.* Ottawa Co., *Kauffman.* Washtenaw Co., *Steere.*

NECKERACEAE

Primary stems inconspicuous, creeping, defoliate; secondary stems erect, horizontal, or pendent, usually strongly flattened, simple or branched; leaves of various shapes, mostly complanate, often undulate; costa lacking, short and double, or single and sometimes well developed; sporophytes lateral on secondary stems; capsules immersed or emergent, rarely exserted, mostly erect and symmetric; peristome single or double; calyptra cucullate, naked or hairy.

Leaves undulate . . . . 1. *Neckera*
Leaves smooth . . . . 2. *Homalia*

1. *Neckera* Hedw.

Secondary stems erect to pendent, ± pinnately to bipinnately branched; leaves complanate, mostly undulate, ovate-lanceolate to

[123]

oblong or lingulate, obtuse to acute or broadly acuminate, ± asymmetric; costa various; cells smooth, short above, becoming linear below, usually yellowish and porose at insertion; capsules immersed to exserted, erect and symmetric, oblong-ovoid; peristome double.

1. *Neckera pennata* Hedw. — Light-green or yellowish, shiny plants with horizontally spreading, freely branched secondary stems; leaves ovate-lanceolate, acute or acuminate, strongly undulate; costa variable; capsules immersed; spores in spring or summer. — On bark of trees, rarely on rock. Widespread in eastern North America; Europe, Canary Islands, Africa, Asia, and the Antipodes.

ILLUSTRATIONS: Figure 98. Conard, *How to Know the Mosses* (ed. 2), fig. 300. Grout, *Mosses with Hand-Lens and Microscope*, fig. 215. Welch, *Mosses of Indiana*, fig. 238.

Upper Peninsula: Alger Co., *Wheeler*. Baraga Co., *Steere*. Chippewa Co., *Thorpe*. Houghton Co., *Parmelee*. Keweenaw Co., *Gilly & Parmelee*. Marquette Co., *Nichols*. Ontonagon Co., *Darlington*.
Lower Peninsula: Alpena Co., *Robinson & Wells*. Cheboygan and Emmet counties, *Wheeler*. Leelanau Co., *Darlington*. Muskegon Co., *Steere*. Van Buren Co., *Kauffman*.

1a. *Neckera pennata* var. *oligocarpa* (Bruch) Grout — Plants more slender, with shorter pointed leaves (broadly obtuse and apiculate) and somewhat shorter upper cells; capsules emergent or laterally exserted. — Mostly on rocks. Canada and far southward in eastern North America, Rocky Mts. to Colorado and Arizona; Europe and Asia.

ILLUSTRATION: Grout, *Mosses with Hand-Lens and Microscope*, fig. 216.

Upper Peninsula: Keweenaw Co., *Steere*.
Lower Peninsula: Cheboygan Co., *Nichols*.

2. *Homalia* (Brid.) BSG

Shiny, green or golden-green plants in low, flat mats, irregularly branched; leaves complanate, not undulate, asymmetric, oblong-lingulate, rounded-obtuse, often inflexed at the base on one side; costa single, slender, ending near the leaf middle; cells rhombic, shorter at apex, longer at base; setae long and slender; capsules erect or inclined, symmetric or nearly so; annulus present; operculum rostrate; peristome double, the teeth transversely striolate; endostome with perforate segments from a high basal membrane.

Homalia jamesii Schimp. — Plants shiny, yellow-green, strongly complanate-foliate; setae up to 1.5 mm. long; capsules erect; spores in autumn. — On moist, shaded cliffs, sometimes on tree bases.

[124]

Newfoundland to British Columbia, south to Pennsylvania and Washington.

ILLUSTRATIONS: Figure 99. Conard, *How to Know the Mosses* (ed. 2), fig. 223.

Upper Peninsula: Alger Co., *Nichols.* Chippewa and Keweenaw counties, *Steere.* Marquette Co., *Nichols.* Ontonagon Co., *Nichols* & *Steere.*
Lower Peninsula: Cheboygan Co., *Nichols* & *Steere.* Emmet Co., *Steere.*

THELIACEAE

Plants creeping (or crowded and appearing erect or ascending), in loose or dense, often glaucous mats, sparsely branched to closely pinnate; leaves crowded and imbricate, concave, broadly ovate to suborbicular, rounded-obtuse to acuminate; costa single, short and double, or nearly lacking; cells rhombic, mostly stoutly unipapillose, quadrate at basal angles; setae elongate; capsules erect to horizontal, symmetric or nearly so; operculum conic; peristome double, the endostome sometimes poorly developed; calyptra cucullate.

Paraphyllia lacking; segments well developed;
  cilia present . . . . 1. *Myurella*
Paraphyllia present; segments of endostome rudimentary or
  lacking; cilia none . . . . 2. *Thelia*

1. *Myurella* BSG

Slender, julaceous plants in loose, bluish or glaucous mats or dense tufts, prostrate or erect-ascending (when crowded), irregularly branched; leaves rounded-ovate, rounded-obtuse, sometimes abruptly apiculate or acuminate, very concave; costa very short and double or nearly lacking; cells elliptic to rhombic, bearing a central papilla or sometimes ± papillose at back because of projecting ends; dioicous; capsules exserted, erect to inclined, oblong-ovoid; peristome perfect, the endostome with a basal membrane, segments, and cilia.

Leaves crowded, rounded-obtuse, sometimes ± apiculate;
  cells papillose because of projecting angles . . . 1. *M. julacea*
Leaves rather distant, rather abruptly acuminate;
  papillae central . . . . 2. *M. sibirica*

1. *Myurella julacea* (Schwaegr.) BSG — Plants small, julaceous, grayish-green to glaucous or bluish, often crowded and suberect; leaves crowded, imbricate, rounded-obtuse, sometimes ± apiculate, irregularly serrulate and ± papillose at back because of projecting angles; spores in spring or summer. — In crevices of calcareous

[125]

cliffs, sometimes on sheltered, peaty banks. Greenland; across the continent, south to New York, Colorado, and Oregon.

ILLUSTRATIONS: Figure 100. Grout, *Moss Flora of North America*, vol. 3, Pl. 50.

Upper Peninsula: Keweenaw Co., *Steere*. Marquette Co., *Nichols*. Ontonagon Co., *Nichols* & *Steere*.

Lower Peninsula: Cheboygan Co., *Nichols*. Emmet Co., *Steere*. Huron Co., *Schnooberger*. Presque Isle Co., *Steere*.

2. *Myurella sibirica* (C.M.) Reim. — Plants slender, light-green or glaucous, terete-foliate; leaves rather distant, loosely imbricate or erect, rather abruptly acuminate; margins spinose-serrate; cells coarsely unipapillose at back, the papilla central. — In crevices or hollows of calcareous cliffs. Widespread in eastern North America; Europe and Asia.

ILLUSTRATIONS: Conard, *How to Know the Mosses* (ed. 2), fig. 176 (as *M. careyana*). Grout, *Moss Flora of North America*, vol. 3, Pl. 51 (as *M. careyana*). Welch, *Mosses of Indiana*, fig. 229 (as *M. careyana*).

Upper Peninsula: Keweenaw Co., *Steere*. Mackinac Co., *Nichols*. Marquette Co., *Hill*. Ontonagon Co., *Steere*.

Lower Peninsula: Alpena Co., *Robinson* & *Wells*. Cheboygan Co., *Steere*.

2. *Thelia* Sull.

Yellowish, dull-green, or glaucous plants in low mats, pinnately branched; stems creeping; branches short and julaceous, erect or ascending; paraphyllia present; leaves imbricate, concave, rounded-ovate, abruptly acuminate; margins erect, mostly ciliate-serrate, ciliate to laciniate below; costa single and reaching the leaf middle or short, often double; cells rhombic to fusiform, coarsely unipapillose; dioicous; setae elongate; capsules oblong-cylindric, erect and symmetric; peristome double, the endostome consisting of a basal membrane, the segments rudimentary or lacking.

Thelia asprella Sull. — Leaves with ciliate margins; cells bearing 2–3-branched papillae; spores in autumn. — On the bases of trees. Widespread in eastern North America.

ILLUSTRATIONS: Figure 101. Conard, *How to Know the Mosses* (ed. 2), fig. 178a–d. Jennings, *Mosses of Western Pennsylvania* (ed. 2), Pl. 38.

Lower Peninsula: Cheboygan Co., *Roberts*. Monroe Co., *Steere*. Oakland Co., *Darlington*. Washtenaw Co., *Kauffman*.

FABRONIACEAE[*]

Plants small or even minute, creeping, freely branched; leaves erect or imbricate when dry, ovate or lanceolate, acuminate; costa single, extending to the midleaf, or lacking; cells rhombic or oblong-hexagonal, quadrate toward the basal margins, mostly smooth, thin-walled; setae elongate; capsules erect or slightly inclined, symmetric; annulus present; peristome usually present, single or double; operculum conic to short-rostrate; calyptra cucullate, naked.

Anacamptodon Brid.

Plants rather small, dark-green; leaves ovate-lanceolate, gradually acuminate, somewhat concave; margins plane and entire; costa ending at mid-leaf or beyond; cells rhombic-hexagonal, about 3–5:1, a few at the basal angles subquadrate; capsules erect and symmetric, oblong-ovoid, strongly contracted below the mouth when dry and empty; peristome double with the teeth united in pairs and recurved when dry, the segments filiform, shorter than the teeth, basal membrane none; operculum short-rostrate. — We have a single species, A. *splachnoides* (Froel.) Brid. — On bark of trees, particularly in knotholes and fissures. Rare but widespread in eastern North America; Europe.

ILLUSTRATIONS: Figure 102. Grout, *Mosses with Hand-Lens and Microscope*, fig. 210. Jennings, *Mosses of Western Pennsylvania* (ed. 2), Pl. 37. Welch, *Mosses of Indiana*, fig. 242.

Lower Peninsula: Cheboygan Co., *Phillips*. Emmet Co., *Steere*.

LESKEACEAE

Plants small to fairly large, creeping, in loose or dense, dull mats; stems freely branched; stem and branch leaves similar; paraphyllia sometimes present; leaves ovate or ovate-lanceolate; costa single, usually well developed, rarely lacking or short and double; cells firm, often incrassate, smooth or papillose; setae lateral, elongate; capsules erect and symmetric to curved and inclined; peristome double, the endostome ± imperfect; calyptra cucullate.

1. Costa poorly developed (single or double,
   very short) . . . . 4. *Pterigynandrum*
1. Costa single, reaching at least the leaf middle 2
   2. Leaves with strongly recurved margins; capsules

[*]Mazzer and Sharp (Bryol. 66: 68-69. 1963) have recently reported *Fabronia ciliaris* (Brid.) Brid. from Ionia and Van Buren counties. This extremely small and delicate pleurocarp is easily recognized by its ovate-lanceolate, slenderly acuminate, irregularly dentate leaves, weak costa (ending near the leaf middle), single peristome, and exothecial cells with sinuose walls. It is found on rocks and bark of trees almost throughout the eastern U. S., Arizona to Guatemala, and Europe.

inclined and asymmetric; segments of endostome
lanceolate; cilia usually present . . . . 1. *Pseudoleskea*
2. Leaves not or slightly recurved at margins; capsules
erect and symmetric; segments linear or lacking; cilia
rudimentary or lacking 3
3. Leaves squarrose when moist; cells stoutly unipapillose;
peristome teeth blunt; endostome segments
lacking . . . . 3. *Lindbergia*
3. Leaves erect-spreading when moist; cells smooth or
bluntly and obscurely papillose; peristome teeth slenderly
pointed; segments present . . . . 2. *Leskea*

1. *Pseudoleskea* BSG

Plants rather small to medium-sized, in loose, dull mats; stems freely branched, ± pinnate; branches usually recurved at the tips; paraphyllia present, sometimes abundant; leaves symmetric or curved, ovate-lanceolate, acuminate, often biplicate; margins recurved below, usually denticulate or serrulate above; costa well developed; cells subquadrate to oblong-rhomboidal, often unipapillose; setae elongate; capsules inclined and curved; peristome double, the endostome consisting of basal membrane and well-developed segments, the cilia lacking or rudimentary; calyptra cucullate.

Median cells of stem leaves at least 3–4:1; cells of branch
leaves usually papillose at the upper ends at back; paraphyllia
multiform; leaves loosely appressed when dry . . 2. *P. radicosa*
Median cells about 2:1 or less; papillae on both surfaces,
central; paraphyllia lanceolate; leaves closely
appressed when dry . . . . 1. *P. patens*

1. *Pseudoleskea patens* (Lindb.) Limpr. — Plants loosely spreading, irregularly pinnate, radiculose; paraphyllia lanceolate; leaves ovate, obliquely acuminate, lightly biplicate, costate into the acumen; cells short, with central papillae; spores in spring. — On rocks. Alaska to California and Wyoming, Michigan, New England and southeastern Canada; Europe and Iceland.

ILLUSTRATIONS: Figure 103. Conard, *How to Know the Mosses* (ed .2), fig. 180. Grout, *Moss Flora of North America*, vol. 3, Pl. 53 (fig. 1–7).

Upper Peninsula: Keweenaw Co., *Steere*.

2. *Pseudoleskea radicosa* (Mitt.) Macoun & Kindb. — Plants in rather soft, radiculose mats, irregularly branched; paraphyllia numerous, multiform; stem leaves secund, contracted from an ovate or oblong-ovate base to a narrow, oblique acumen, lightly biplicate; costa subpercurrent; cells elongate, papillose at the upper ends at back; spores in spring and early summer. (Michigan reports of *P. oligoclada* belong here, according to Lawton) — On rocks, rarely

on soil. Alaska to Colorado and Arizona; rare in eastern Canada, New Hampshire, and Michigan; Europe and Asia.

ILLUSTRATION: Grout, *Moss Flora of North America*, vol. 3, Pl. 54 (fig. 1–9).

Upper Peninsula: Keweenaw Co., *Steere*. Ontonagon Co., *Nichols & Steere*.

2. *Leskea* Hedw.

Small plants in low, dull, usually dark-green mats; stems prostrate or ascending; leaves crowded, small, ovate or ovate-lanceolate, often asymmetric, usually entire; costa mostly single (rarely very short and double); cells short, mostly obscurely papillose; setae elongate; capsules cylindric, erect, usually somewhat curved; peristome double, basal membrane of endostome low, segments linear, cilia rudimentary or lacking; calyptra cucullate.

1. Median leaf cells elongate 2
1. Median cells isodiametric 3
  2. Leaves abruptly narrowed to a lance-linear acumen; costa extending into the acumen . . . . . 1. *L. nervosa*
  2. Leaves shortly acuminate; costa ending below the acumen . . . . 2. *L. tectorum*
3. Leaves more than 2:1 4
3. Leaves less than 2:1 5
  4. Capsules straight; operculum short-conic . . 3. *L. polycarpa*
  4. Capsules curved; operculum long-conic (nearly ½ as long as the urn) . . . . 4. *L. arenicola*
5. Leaves symmetric, lightly biplicate; margins often revolute . . . . 5. *L. gracilescens*
5. Leaves asymmetric, not plicate; margins plane . . 6. *L. obscura*

1. *Leskea nervosa* (Schwaegr.) Myr. — Plants in thin, dark-green or blackish mats; leaves small, ovate, abruptly narrowed to a lance-linear, often recurved subula; costa ending in the acumen; cells somewhat longer than wide; spores in summer. — On calcareous rocks, sometimes on bases of trees. Across Canada, south to Pennsylvania and Colorado; Europe and Asia.

ILLUSTRATIONS: Conard, *How to Know the Mosses* (ed. 2), fig. 187. Grout, *Mosses with Hand-Lens and Microscope*, Pl. 57.

Upper Peninsula: Houghton Co., *Parmelee*. Keweenaw Co., *Hermann*. Luce and Marquette counties, *Nichols*. Ontonagon Co., *Darlington & Bessey*.
Lower Peninsula: Alpena Co., *Robinson & Wells*. Cheboygan Co., *Nichols*. Kalamazoo Co., *Becker*. Leelanau Co., *Darlington*.

1a. *Leskea nervosa* var. *nigrescens* (Kindb.) Best — Plants smaller and sterile, bearing numerous flagelliform branchlets with rudimentary leaves. — Ranging with the species in North America.

Lower Peninsula: Leelanau Co. (Glen L.), *Darlington*.

2. *Leskea tectorum* (A. Br.) Lindb. — Plants deep-green to red-brown; branches ascending; leaves ovate, abruptly short-acuminate; costa ending below the acumen; cells somewhat longer than broad; spores in late summer. — On rocks (usually calcareous), rarely on bases of trees. Greenland; Yukon to the southern Rocky Mts.; Lake Superior; Europe and Asia.

ILLUSTRATION: Grout, *Moss Flora of North America*, vol. 3, Pl. 52 (fig. 69–76).

Upper Peninsula: Keweenaw Co., *Hermann*. Marquette Co., *Nichols*. Ontonagon Co., *Nichols* & *Steere*.

3. *Leskea polycarpa* Hedw. — Plants pale- to dark-green; leaves oblong-ovate, gradually acuminate and blunt or acute, ± secund and obliquely pointed; costa disappearing below the apex; cells isodiametric, with 1–2 small papillae on either surface; spores in early summer. — On bark at base of trees, sometimes on rocks. Newfoundland and southeastern Canada, south to the central United States; Europe and Asia.

ILLUSTRATIONS: Figure 104. Conard, *How to Know the Mosses* (ed. 2), fig. 188e–f. Grout, *Mosses with Hand-Lens and Microscope*, Pl. 56.

Upper Peninsula: Ontonagon Co., *Darlington* & *Bessey*.
Lower Peninsula: Clinton Co., *Parmelee*. Macomb Co., *Cooley*. Washtenaw Co., *Steere*.

4. *Leskea arenicola* Best — Plants yellowish-green to red-brown; leaves ovate or ovate-lanceolate, obliquely acuminate; costa disappearing near the apex; cells isodiametric or somewhat elongate, unipapillose; spores in early summer. — On tree bases, rarely on rotten wood, in sandy places. New England to Virginia, west to North Dakota.

ILLUSTRATIONS: Conard, *How to Know the Mosses* (ed. 2), fig. 188g. Grout, *Mosses with Hand-Lens and Microscope*, Pl. 57.

Lower Peninsula: Grand Traverse Co. (Sucker Creek), *R. Conard*.

5. *Leskea gracilescens* Hedw. — Plants usually dark-green or blackish; leaves symmetric, ovate, obtuse to acute, lightly biplicate; margins often recurved; cells isodiametric, unipapillose on the lower surface; spores in early summer. — On the bases of trees, rotten wood, soil, and stones; southeastern Canada and northeastern United States.

[130]

ILLUSTRATIONS: Conard, *How to Know the Mosses* (ed. 2), fig 188a–c. Grout, *Moss Flora of North America*, vol. 3, Pl. 71B. Welch, *Mosses of Indiana*, fig. 226.

Upper Peninsula: Keweenaw Co., *Hermann.*
Lower Peninsula: Berrien Co., *Hill.* Eaton Co., *Parmelee.* Ingham Co., *Darlington.*

6. *Leskea obscura* Hedw. — Plants dark-green; leaves asymmetric, oblong-ovate, obtuse or subacute, scarcely plicate; margins plane; costa disappearing below the apex; cells isodiametric, unipapillose on the lower and sometimes also on the upper surface; spores in early summer. — On the bases of trees or rotten wood, rarely on soil or stones. Widespread in eastern North America.

ILLUSTRATIONS: Conard, *How to Know the Mosses* (ed. 2), fig. 188d. Grout, *Mosses with Hand-Lens and Microscope*, Pl. 57 (fig. 14–27). Jennings, *Mosses of Western Pennsylvania* (ed. 2), Pl. 39, 67. Welch, *Mosses of Indiana*, fig. 227.

Lower Peninsula: Gratiot Co., *Schnooberger.* Ingham Co., *Darlington.* Van Buren Co., *Kauffman.*

3. *Lindbergia* Kindb.

Small, dark-green or blackish plants; leaves ovate or ovate-lanceolate, gradually or abruptly acuminate, entire or nearly so, squarrose-spreading when moist; costa single, ending at or beyond the leaf middle; cells rhombic, unipapillose; setae elongate; capsules erect and symmetric (or weakly curved); peristome double, the teeth bluntly lanceolate, the endostome consisting of a low basal membrane.

Lindbergia brachyptera (Mitt.) Kindb. — Plants often bearing clusters of small branches in leaf axils; leaves ovate, abruptly narrowed to a slender acumen. — On bark, rarely on rocks or logs. Widespread in eastern North America; Arizona and West Texas; Caucasus.

ILLUSTRATIONS: Figure 105. Conard, *How to Know the Mosses* (ed. 2), fig. 181 (as var. *austinii*).

Upper Peninsula: Ontonagon Co., *Nichols & Steere.*
Lower Peninsula: Cheboygan Co., *Nichols.* Emmet Co., *Steere.* Kalamazoo Co., *Becker.* Presque Isle and Washtenaw counties, *Steere.*

Pterigynandrum Hedw.

Rather small, yellowish or green plants in loose or dense mats; stems creeping; branches ascending and curved, slender and terete; leaves crowded, imbricate or slightly secund, concave, oblong-ovate, acute to acuminate; margins narrowly reflexed below; costa mostly very short and thin, double or forked, sometimes single and reaching the leaf middle; cells rhomboid-hexagonal, mostly papillose at back because of projecting ends; setae elongate; capsules cylindric, erect and symmetric; peristome double, the endostome without cilia; calyptra cucullate.

Pterigynandrum filiforme Hedw. — Small, dark- or yellow-green mosses in loose mats, often mixed with other mosses; leaves slightly secund; costa short; cells clearly papillose at back. — On bases of trees or on boulders. Greenland; across Canada and northern United States; Europe, North Africa, Canary Islands, and Asia.

ILLUSTRATIONS: Figure 106. Grout, *Mosses with Hand-Lens and Microscope*, fig. 134.

Upper Peninsula: Keweenaw Co., *Steere*. Marquette Co., *Nichols*. Ontonagon Co., *Nichols* & *Steere*.
Lower Peninsula: Cheboygan Co., *Nichols*. Leelanau Co., *Darlington*.

THUIDIACEAE

Small to large, often handsome plants in loose mats or tufts; stems prostrate to erect or ascending, freely branched, sometimes regularly pinnate and often frondose; paraphyllia often present; stem and branch leaves often differentiated; costa single, well developed; cells firm or thick-walled, mostly short, often papillose; setae elongate; capsules erect or inclined, often asymmetric; peristome double, mostly well developed.

1. Stem and branch leaves similar . . . . . 4. *Anomodon*
1. Stem and branch leaves differentiated in size and shape 2
  2. Paraphyllia none or very sparse . . . . 3. *Heterocladium*
  2. Paraphyllia abundant 3
    3. Leaf cells isodiametric . . . . 1. *Thuidium*
    3. Leaf cells oblong-linear . . . . 2. *Helodium*

1. *Thuidium* BSG

Small to robust plants in loose mats or cushions; stems creeping to erect-ascending, usually regularly 1–3-pinnate, often forming triangular fronds suggestive of a fern; paraphyllia abundant, papillose, polymorphous; stems and branch leaves differentiated; stem leaves ovate, acuminate, generally ± plicate; branch leaves small, concave, rounded-obtuse to acute; costa well-developed; cells coarsely unipapillose or finely pluripapillose; setae elongate; capsules inclined to horizontal, ± curved; peristome teeth cross-striolate below, endostome consisting of a high basal membrane, keeled segments, and (usually) well-developed cilia; calyptra cucullate, usually naked.

[132]

1. Plants irregularly or subpinnately branched; apical cell
of branch leaves with a single, terminal papilla 2
1. Plants regularly 1–2-pinnate; apical cell of branch
leaves with 2–4 papillae 3
  2. Stem leaves abruptly short-acuminate, erose-serrate;
  operculum short-rostrate . . . . . . 8. *T. virginianum*
  2. Stem leaves gradually long-acuminate, crenulate-
  serrulate; operculum obtuse or
  conic-apiculate . . . . 7. *T. microphyllum*
3. Plants rigid, 1-pinnate 4
3. Plants softer, 2-pinnate 5
  4. Leaf cells unipapillose . . . . . . 4. *T. abietinum*
  4. Leaf cells pluripapillose . . . . 6. *T. scitum*
5. Small plants; cells pluripapillose . . . . . 5. *T. minutulum*
5. Rather large plants; cells unipapillose 6
  6. Stem leaves incurved, with abruptly spreading
  tips, costa filling the acumina, and
  margins plane or nearly so . . . . . 3. *T. recognitum*
  6. Stem leaves loosely erect, with costa ending well
  below the apex and the margins recurved 7
7. Stem leaves ending in a hyaline, filiform tip;
perichaetial leaves not ciliate . . . . . . 2. *T. philibertii*
7. Stem leaves acute or acuminate; perichaetial
leaves ciliate . . . . 1. *T. delicatulum*

1. *Thuidium delicatulum* (Hedw.) BSG — Plants rather large,
attractively frondose, bipinnately branched, green or yellow-green;
stem leaves ovate, broadly acuminate, plicate, with the costa ending
in the apex; branch leaves small, acute, ending in a 2–4-papillose
cell; cells stoutly unipapillose; perichaetial leaves ciliate; spores in
late autumn or winter. — On various substrata in moist or wet
woods. Widespread in eastern North America; Europe, Asia, Central
and South America.

ILLUSTRATIONS: Figure 107. Conard, *How to Know the Mosses* (ed. 2), fig.
193, 194c. Grout, *Mosses with Hand-Lens and Microscope*, fig. 128, 133a.
Jennings, *Mosses of Western Pennsylvania* (ed. 2), Pl. 40. Welch, *Mosses of
Indiana*, fig. 217.

Upper Peninsula: Chippewa Co., *Thorpe.* Keweenaw Co., *Mugford.* Mar-
quette Co., *Nichols.* Ontonagon Co., *Darlington & Bessey.*
Lower Peninsula: Alpena Co., *Hall.* Cheboygan Co., *Nichols.* Clinton Co.,
Parmelee. Ingham, Leelanau, Macomb, and Oakland counties, *Darlington.*
Washtenaw Co., *Kauffman.*

2. *Thuidium philibertii* Limpr. — Very similar to *T. delicatulum;*
stem leaves longer, ending in a hyaline tip of 2–9 cells; perichaetial
leaves not ciliate. — On soil, humus, or logs in moist woods. Ontario
to Alaska and the Aleutian Islands, south to Alberta and the eastern
United States to Virginia; Europe and Asia.

ILLUSTRATION: Grout, *Moss Flora of North America*, vol. 3, Pl. 44A.

Upper Peninsula: Keweenaw Co., *Allen & Stuntz.* Ontonagon Co., *Nichols & Steere.*

Lower Peninsula: Montcalm and Otsego counties, *Schnooberger.*

3. *Thuidium recognitum* (Hedw.) Lindb. — Differing from *T. delicatulum* in the stem leaves which are concave and incurved, with abruptly spreading tips filled by the costa; margins plane; perichaetial leaves not ciliate. — On various substrata in moist woods. Alaska to Newfoundland, south to Oregon, Arkansas, and Georgia; Europe and Asia.

ILLUSTRATIONS: Conard, *How to Know the Mosses* (ed. 2), fig. 194a–b. Grout, *Mosses with Hand-Lens and Microscope*, Pl. 52, fig. 8 and 127. Jennings, *Mosses of Western Pennsylvania* (ed. 2), Pl. 40. Welch, *Mosses of Indiana*, fig. 219.

Upper Peninsula: Chippewa Co., *Steere.* Keweenaw Co., *Hermann.* Marquette Co., *Nichols.*

Lower Peninsula: Alpena Co., *Wheeler.* Cheboygan Co., *Nichols.* Leelanau Co., *Darlington.* Shiawasee Co., *Marshall.* Washtenaw Co., *Steere.*

4. *Thuidium abietinum* (Hedw.) BSG — Rather large, brownish-green, rigid plants; stems erect or ascending, 1-pinnate; cells bluntly unipapillose; spores in spring. — On rocks, sandy soil, or turf in dry, usually limey habitats. Greenland; across the continent, south to Colorado and Virginia; Europe and Asia.

ILLUSTRATIONS: Conard, *How to Know the Mosses* (ed. 2), fig. 192a–d. Grout, *Mosses with Hand-Lens and Microscope*, fig. 126b, 133d, *Moss Flora of North America*, vol. 3, Pl. 37. Welch, *Mosses of Indiana*, fig. 220.

Upper Peninsula: Keweenaw Co., *Parmelee.* Mackinac Co., *Wheeler.* Marquette Co., *Nichols.* Ontonagon Co., *Darlington.*

Lower Peninsula: Alpena and Cheboygan counties, *Wheeler.* Emmet Co., *Nichols.*

5. *Thuidium minutulum* (Hedw.) BSG — Small, dark-green or yellow-green plants, usually 2-pinnate; cells pluripapillose; spores in autumn. — On rotten wood, rocks, or soil, etc., in moist woods. Widespread in eastern North America; tropical America, Azores; Europe.

ILLUSTRATIONS: Conard, *How to Know the Mosses* (ed. 2), fig. 191a–c. Grout, *Mosses with Hand-Lens and Microscope*, fig. 122a, Pl. 52 (fig. 2), *Moss Flora of North America*, vol. 3, Pl. 37. Jennings, *Mosses of Western Pennsylvania* (ed. 2), Pl. 39. Welch, *Mosses of Indiana*, fig. 221.

Lower Peninsula: Cheboygan and Ingham counties, *Steere.* Leelanau Co., *Darlington.* Macomb Co., *Cooley.* Washtenaw Co., *Steere.*

6. *Thuidium scitum* (P.-Beauv.) Aust. — Small, rigid, dark-green or brownish plants, closely 1-pinnate; branches terete; cells pluri-papillose; spores in autumn or winter. — On bark of trees, especially at the base, and sometimes on logs, stumps, or soil in woods. Widespread in eastern North America.

ILLUSTRATIONS: Conard, *How to Know the Mosses* (ed. 2), fig. 192e. Grout, *Mosses with Hand-Lens and Microscope,* fig. 126a, 133b, *Moss Flora of North America,* vol. 3, Pl. 44B. Jennings, *Mosses of Western Pennsylvania* (ed. 2), Pl. 39 (as *Rauia*).

Lower Peninsula: Cheboygan Co., *Nichols.*

7. *Thuidium microphyllum* (Hedw.) Best — Plants in loose, light-green or yellowish mats; stems irregularly or subpinnately branched; paraphyllia few to numerous; stem leaves long-acuminate, crenulate-serrulate; cells unipapillose; terminal cell of branch leaves unipapillose; spores in summer. — On various substrata, often in rather dry, brushy, disturbed places. Eastern North America; Mexico to South America; Europe and Asia.

ILLUSTRATIONS: Grout, *Mosses with Hand-Lens and Microscope,* Pl. 55. Jennings, *Mosses of Western Pennsylvania* (ed. 2), Pl. 39 (as *Haplocladium*). Welch, *Mosses of Indiana,* fig. 223.

Upper Peninsula: Gogebic Co., *Bessey.* Houghton and Keweenaw counties, *Parmelee.* Ontonagon Co., *Darlington & Bessey.*
Lower Peninsula: Cheboygan Co., *Nichols.*

8. *Thuidium virginianum* (Brid.) Lindb. — Plants subpinnately branched, dark-green; stem leaves abruptly short-acuminate, erose-dentate; cells unipapillose; terminal cell of branch leaves unipapillose; spores in spring. — On soil or roots of trees in open woods. Widespread in eastern North America; Europe and Asia.

ILLUSTRATIONS: Conard, *How to Know the Mosses* (ed. 2), fig. 190, 194d. Grout, *Mosses with Hand-Lens and Microscope,* Pl. 54. Jennings, *Mosses of Western Pennsylvania* (ed. 2), Pl. 39 (as *Haplocladium*). Welch, *Mosses of Indiana,* fig. 224.

Upper Peninsula: Houghton Co., *Parmelee.* Ontonagon Co., *Darlington.*
Lower Peninsula: Huron Co., *Kauffman.* Washtenaw Co., *Steere.*

2. *Helodium* (Sull.) Warnst.

Plants in soft tufts; stems suberect, freely to pinnately branched; branches slender, tapered; paraphyllia very numerous, filiform, branched; leaves erect or appressed when dry, erect-spreading when moist, broadly ovate to ovate-lanceolate, sharply pointed, concave, plicate; margins partially recurved; costa ending below the apex to nearly percurrent; cells elongate, smooth or unipapillose; branch leaves smaller; setae elongate; capsules curved, inclined to horizontal; annulus differentiated; peristome double, complete.

Plants regularly 1-pinnate; costa ending above the leaf
  middle; cells with a large papilla at the upper ends .   1. *H. blandowii*
Plants irregularly pinnate; costa nearly percurrent; cells
  smooth or with a small papilla at the upper ends .   2. *H. paludosum*

1. *Helodium blandowii* (Web. & Mohr) Warnst. — Soft, light-green or yellowish, 1-pinnate plants with tomentose stems; leaves broadly ovate; costa vanishing above the leaf middle; median cells 3–6:1, unipapillose near the upper ends; autoicous; spores in summer. — On soil or humus in swamps and wet woods, often at the margins of lakes or ponds. Greenland; across the continent and south to Ohio and Colorado; Europe and Asia.

ILLUSTRATIONS: Figure 108. Grout, *Mosses with Hand-Lens and Microscope*, Pl. 52 (fig. 4–5), fig. 132.

Upper Peninsula: Keweenaw Co., *Allen & Stuntz.*
Lower Peninsula: Cheboygan Co., *Nichols.* Washtenaw Co., *Steere.*

2. *Helodium paludosum* (Sull.) Aust. — Plants irregularly pinnate, the branches of unequal length; leaves oblong- to ovate-lanceolate; costa subpercurrent; cells smooth or minutely unipapillose at the upper ends; autoicous; spores in May. — On soil in swamps and wet meadows. Widespread in eastern North America; Asia.

ILLUSTRATIONS: Conard, *How to Know the Mosses* (ed. 2), fig. 189. Grout, *Mosses with Hand-Lens and Microscope*, Pl. 53. Jennings, *Mosses of Western Pennsylvania* (ed. 2), Pl. 40. Welch, *Mosses of Indiana*, fig. 225.

Lower Peninsula: Clinton Co., *Parmelee.* Leelanau Co., *Darlington.* Washtenaw Co., *Steere.*

3. *Heterocladium* BSG

Small to medium-sized plants in loose or dense mats; stems irregularly or subpinnately branched; paraphyllia few; stem and branch leaves differentiated; stem leaves triangular-ovate or cordate and clasping at base, narrowly acuminate; branch leaves smaller, narrower, ovate to ovate-lanceolate, obtuse or acute to short-acuminate; costa short, double or forking, rarely single and ending near the leaf middle; cells ± elongate in the median region, shorter toward the margins, with 1 or more papillae per cell; dioicous; perichaetial leaves sheathing; setae elongate; capsules curved, inclined to horizontal; annulus present; peristome double, complete.

Heterocladium squarrosulum (Voit) Lindb. — Stem leaves triangular-ovate and clasping at base, abruptly narrowed to a slender, recurved acumen; costa usually short, double or forked, sometimes reaching the leaf middle, sometimes obsolete; ultimate branch leaves rounded-ovate, rounded-obtuse to subacute; spores in late winter or spring. — On soil, humus, or rocks in moist woods. Greenland; British Columbia to Montana and Washington; Labrador to Vermont and Michigan; Europe; Caucasus; Ecuador.

ILLUSTRATIONS: Figure 109. Grout, *Moss Flora of North America*, vol. 3, Pl. 48 (fig. 18–40).

Upper Peninsula: Houghton Co., *Parmelee*. Keweenaw Co., *Hermann*. Luce and Marquette counties, *Nichols*. Ontonagon Co., *Darlington*.
Lower Peninsula: Cheboygan and Emmet counties, *Nichols*.

4. *Anomodon* Hook. & Tayl.

Mostly rather robust plants in dense tufts, dark-green or brownish; stems irregularly branched; paraphyllia none; stem and branch leaves similar; leaves lanceolate to lingulate, rounded-obtuse to apiculate or acuminate; margins plane and entire; costa strong, mostly ending below the apex; cells small, rounded-hexagonal, densely papillose, only the median basal cells elongate and smooth; dioicous; setae elongate; capsules mostly erect and symmetric; peristome double, the endostome with rather short segments from a low basal membrane, cilia lacking or rudimentary; calyptra cucullate.

1. Branches in part slenderly tapered
   to flagelliform . . . . 1. *A. attenuatus*
1. Branches not attenuate 2
   2. Branches terete and almost filiform when dry;
      leaf points often broken off . . . . 2. *A. tristis*
   2. Branches coarser, terete or somewhat flattened;
      leaves not fragile 3
3. Branches terete; leaves slenderly acuminate
   and hair-pointed . . . . 3. *A. rostratus*
3. Branches not terete, ± flattened; leaves obtuse
   or broadly pointed 4
   4. Leaves lanceolate and tapered from an ovate
      base to a subacute apex . . . . . . 4. *A. viticulosus*
   4. Leaves lingulate and obtuse or rounded-obtuse,
      not tapered to the apex 5
5. Leaves not auriculate, rounded-obtuse, entire and
   not apiculate, erect and scarcely contorted when dry . 5. *A. minor*
5. Leaves auriculate, obtuse and often apiculate or serrulate
   at apex, ± crisped and incurved when dry . . . 6. *A. rugelii*

1. *Anomodon attenuatus* (Hedw.) Hüb. — Dull, dark-green plants with decurved branches, many of them slenderly attenuate; leaves sublingulate and subacute or rounded-obtuse and apiculate

from a broadly ovate base; costa ending near the apex; spores in autumn. — On bark of trees (especially at the base), rocks, and other substrata, generally in woods. Widespread in eastern North America; Arizona to Guatemala; West Indies; Europe and Asia.

ILLUSTRATIONS: Conard, *How to Know the Mosses* (ed. 2), fig. 186a–c. Grout, *Mosses with Hand-Lens and Microscope*, fig. 137. Jennings, *Mosses of Western Pennsylvania* (ed. 2), Pl. 38. Welch, *Mosses of Indiana*, fig. 234.

Upper Peninsula: Alger Co., *Gilly & Parmelee.* Chippewa Co., *Steere.* Gogebic Co., *Golley.* Keweenaw Co., *Hermann.* Mackinac Co., *Wheeler.* Marquette Co., *Nichols.*

Lower Peninsula: Alpena Co., *Robinson & Wells.* Cheboygan Co., *Nichols.* Eaton Co., *Darlington.* Emmet Co., *Nichols.* Ingham, Leelanau, and Oakland counties, *Darlington.* Washtenaw Co., *Kauffman.*

2. *Anomodon tristis* Sull. — Plants small and rigid, dark- or brownish-green, in thin, loose mats; leaves appressed when dry, squarrose when moist; leaves narrowly lingulate from an ovate base, often apiculate (the points often broken off); costa ending near the leaf middle; not known to fruit in North America. — On bark of trees, occasionally on rocks or logs. Widespread in eastern North America; Mexico; Europe and Asia.

ILLUSTRATIONS: Grout, *Moss Flora of North America*, vol. 3, Pl. 57C. Jennings, *Mosses of Western Pennsylvania* (ed. 2), Pl. 38 (as *Haplohymenium*). Welch, *Mosses of Indiana*, fig. 236.

Upper Peninsula: Gogebic Co., *Conard.* Marquette Co., *Nichols.*
Lower Peninsula: Montcalm Co., *Schnooberger.*

3. *Anomodon rostratus* (Hedw.) Schimp. — Plants in yellow-green or brownish, dense mats; leaves lance-acuminate from an ovate base, ending in a hyaline hair-point; costa ending a little below the apex; spores in autumn. — On rock (especially limestone) and other substrata in moist, shady places. Widespread in eastern North America; Arizona to Guatemala; West Indies; Bermuda; Europe and Asia.

ILLUSTRATIONS: Conard, *How to Know the Mosses* (ed. 2), fig. 183. Grout, *Mosses with Hand-Lens and Microscope*, fig. 138. Jennings, *Mosses of Western Pennsylvania* (ed. 2), Pl. 38. Welch, *Mosses of Indiana*, fig. 235.

Upper Peninsula: Alger Co., *Gilly & Parmelee.* Delta Co., *Wheeler,* Marquette Co., *Nichols.* Ontonagon Co., *Darlington & Bessey.*

Lower Peninsula: Alpena Co., *Robinson & Wells.* Charlevoix Co., *Wheeler.* Cheboygan Co., *Nichols.* Clinton Co., *Parmelee.* Emmet Co., *Nichols.* Ingham, Kalamazoo, Leelanau, and Oakland counties, *Darlington.* Washtenaw Co., *Kauffman.*

4. *Anomodon viticulosus* (Hedw.) Hook. & Tayl. — Coarse, robust, dark-green or brownish plants in dense mats; leaves somewhat contorted when dry, lance-ligulate and broadly tapered from a broader, ovate base to an obtusely acute apex; costa extending nearly to the apex; spores in winter. — On limestone cliffs, also on bark of trees, in moist, shady places. Widespread in eastern North America; Europe and Asia.

ILLUSTRATIONS: Figure 110. Grout, *Mosses with Hand-Lens and Microscope,* fig. 136.

Upper Peninsula: Delta Co., *Steere.* Mackinac Co., *Nichols.*
Lower Peninsula: Alger Co., *Robinson & Wells.* Cheboygan Co., *Nichols.* Emmet Co., *Steere.*

5. *Anomodon minor* (Hedw.) Fürnr. — Dark-green plants in loose mats; leaves broadly lingulate, rounded-obtuse, not apiculate, entire, somewhat decurrent but not auriculate at base, scarcely altered on drying; spores in late fall or winter. — On trunks and bases of trees, logs, and rocks in shade. Widespread in eastern North America; southwestern United States and Mexico; eastern Asia

ILLUSTRATIONS: Conard, *How to Know the Mosses* (ed. 2), fig. 126. Grout, *Mosses with Hand-Lens and Microscope,* Pl. 59. Jennings, *Mosses of Western Pennsylvania* (ed. 2), Pl. 38. Welch, *Mosses of Indiana,* fig. 233.

Upper Peninsula: Alger Co., *Gilly & Parmelee.* Marquette Co., *Nichols.* Ontonagon Co., *Nichols & Steere.*
Lower Peninsula: Alpena Co., *Robinson & Wells.* Cass Co., *Darlington.* Cheboygan Co., *Nichols.* Ingham and Leelanau counties, *Darlington.* Ottawa and Washtenaw counties, *Kauffman.*

6. *Anomodon rugelii* (C.M.) Keissl. — Similar to *A. minor;* leaves incurved-crisped when dry, apiculate or serrulate at apex, not decurrent but with fimbriate-auricles at base; spores in autumn. — On trees and stumps, sometimes on rocks, in shade. Eastern North America; Europe and Asia.

ILLUSTRATIONS: Grout, *Mosses with Hand-Lens and Microscope,* fig. 135a–c (as *A. apiculatus*). Jennings, *Mosses of Western Pennsylvania* (ed. 2), Pl. 38.

Upper Peninsula: Alger, Dickinson, and Mackinac counties, *Wheeler.* Ontonagon Co., *Nichols & Steere.*
Lower Peninsula: Alpena Co., *Robinson & Wells.* Cheboygan and Emmet counties, *Kauffman.*

AMBLYSTEGIACEAE

Plants very small to large, of various habits and habitats but typically found in wet places, sometimes submerged; leaves various

[139]

in shape, usually lanceolate and acuminate, sometimes falcate-secund; costa mostly present, single or double; cells smooth, ± elongate (mostly linear and very long); alar cells frequently inflated; setae elongate; capsules oblong-cylindric, strongly curved and usually horizontal; peristome double and complete (or the cilia sometimes lacking).

1. Paraphyllia abundant, usually ± conspicuous . . 4. *Cratoneuron*
1. Paraphyllia none or few and inconspicuous 2
  2. Leaves with a single costa to the middle or beyond,
    slenderly pointed, ± falcate-secund . . 10. *Drepanocladus*
  2. Leaves not as above (those with a single costa not
    falcate-secund or, if secund, the costa
    double or lacking) 3
3. Costa ending near the apex to percurrent or excurrent 4
3. Costa seldom extending beyond the leaf middle, often
  short and double or lacking 7
  4. Leaves rounded-obtuse, sometimes apiculate . . 8. *Calliergon*
  4. Leaves acuminate 5
5. Leaves acuminate to a narrow, obtuse apex;
  cells linear . . . . 6. *Hygrohypnum*
5. Leaves acuminate and mostly acute; cells oblong-hexagonal
  or oblong-linear 6
  6. Aquatic; costa very strong . . . . 3. *Hygroamblystegium*
  6. Not aquatic; costa slender . . . . . 2. *Amblystegium*
7. Aquatic; leaves mostly obtuse or broadly acute with
  costa double and sometimes short, or lacking 8
7. Not aquatic or, if aquatic, with slenderly acuminate
  leaves and a single costa 9
  8. Plants very large and tumid; leaves very broad,
    subsecund at the tips, rugose when dry; costa
    very faint or lacking . . . . 7. *Scorpidium*
  8. Plants smaller; leaves not rugose;
    costa evident . . . . 6. *Hygrohypnum*
9. Plants small; leaf cells 2–3:1 . . . . . 2. *Amblystegium*
9. Plants medium-sized to large; cells 5–15:1 10
  10. Costa none or short and double . . . 9. *Calliergonella*
  10. Costa single, ending at or beyond
    the leaf middle 11
11. Leaves rounded-obtuse, sometimes apiculate . . 8. *Calliergon*
11. Leaves acuminate and acute 12
  12. Leaves erect-spreading; costa always single
    and well developed . . . . . . 1. *Leptodictyum*
  12. Leaves spreading to squarrose; costa single or
    short and double . . . . 5. *Campylium*

1. *Leptodictyum* (Schimp.) Warnst.

Plants small to fairly robust, in rather loose, low mats in wet places, sometimes in water; leaves spreading or erect-spreading, lanceolate, acute or acuminate; margins plane, entire or nearly so; costa single, reaching the leaf middle or beyond it; cells elongate-hexagonal to linear, thin-walled; setae elongate; capsules strongly curved, horizontal; peristome double, well developed.

1. Rather large plants with stems up to 10 cm. long or
    more and stem leaves 2–2.5 mm. long; median leaf
    cells 10:1 or longer . . . . 1. *L. riparium*
1. Plants smaller; cells short, less than 8:1 2
    2. Leaves ovate-lanceolate; costa ending
        near the apex . . . . 2. *L. trichopodium*
    2. Leaves ovate; costa ending at midleaf
        or somewhat beyond . . . . 3. *L. kochii*

1. *Leptodictyum riparium* (Hedw.) Warnst. — Rather large
plants in soft, light-green, flat mats; stems irregularly branched;
leaves rather distant, lanceolate, gradually and slenderly acuminate,
up to 2.5 mm. long; costa ½–¾ the leaf length; upper cells 8–15:1
above; spores in late spring. — On wet soil, humus, and twigs in
swampy places, sometimes submerged. Widespread in North America; West Indies; Europe, Africa, and Asia.

ILLUSTRATIONS: Figure 111. Grout, *Mosses with Hand-Lens and Microscope*,
fig. 176. Jennings, *Mosses of Western Pennsylvania* (ed. 2), Pl. 42. Welch,
Mosses of Indiana, fig. 155.

Upper Peninsula: Keweenaw Co., *Povah*. Marquette Co., *Nichols*. Ontonagon
Co., *Darlington & Bessey*.
Lower Peninsula: Cheboygan Co., *Nichols*. Clinton Co., *Darlington*. Ingham
Co., *Parmelee*. Leelanau Co., *Darlington*. Washtenaw Co., *Steere*.

2. *Leptodictyum trichopodium* (Schultz) Warnst. — Rather small
plants in loose, low, light-green or yellowish mats; leaves ovate-
lanceolate, rather abruptly acuminate; costa ending near the apex;
upper cells 3–5(8):1; spores in late autumn to early spring. — On
soil and other substrata in swampy places. Widespread in North
America; Europe.

ILLUSTRATIONS: Conard, *How to Know the Mosses* (ed. 2), fig. 217a–e.
Grout, Moss Flora of North America, vol. 3, Pl. 16 (fig. 12). Welch, *Mosses
of Indiana*, fig. 161.

Upper Peninsula: Chippewa Co., *Steere*.
Lower Peninsula: Alpena Co., *Robinson & Wells*. Washtenaw Co., *Steere*.

3. *Leptodictyum kochii* (BSG) Warnst. — Similar to *L. trichopodium;* leaves broader; costa ending at or above the leaf middle;
spores in summer. — On soil and other substrata in swamp habitats.
Across North America; Europe and Asia.

ILLUSTRATIONS: Conard, *How to Know the Mosses* (ed. 2), fig. 217f (as *L.
trichopodium* var.). Grout, *Mosses with Hand-Lens and Microscope*, fig. 173.
Jennings, *Mosses of Western Pennsylvania* (ed. 2), Pl. 42. Welch, *Mosses of
Indiana*, fig. 162 (as *L. trichopodium* var. *kochii*).

Upper Peninsula: Keweenaw Co., *Cooper*. Marquette Co., *Nichols*.
Lower Peninsula: Cheboygan Co., *Nichols*. Ingham and Leelanau counties,
Darlington.

[141]

2. *Amblystegium* BSG

Small, light-green or yellowish, creeping plants growing in moist or rather wet places (but not aquatic); leaves erect to wide-spreading when dry, lanceolate to ovate-lanceolate, gradually acuminate; costa single, slender, ending at or beyond the leaf middle (rarely disappearing in the acumen); cells oblong-hexagonal, up to 5:1; setae elongate; capsules strongly curved, horizontal; peristome double, well developed.

1. Costa vanishing in the narrow acumen to subpercurrent 2
1. Costa ending at or near the leaf middle 3
   2. Leaves entire; cells up to 5:1 . . . . . 3. *A. varium*
   2. Leaves serrulate; cells 6–10:1 . . . . 4. *A. compactum*
   3. Leaves wide-spreading wet or dry; basal
     marginal cells oblong . . . . 2. *A. juratzkanum*
   3. Leaves not wide-spreading; basal marginal
     cells subquadrate . . . . 1. *A. serpens*

1. *Amblystegium serpens* (Hedw.) BSG — Very slender plants; leaves ovate-lanceolate, acuminate, rather close together, not widely spreading; margins entire or serrulate; costa ending at or somewhat above the leaf middle; cells 3–4:1, quadrate or transversely elongate at the basal margins; spores in spring. — On soil, rocks, bark at base of trees, etc. in wet places. Widespread in North America; Europe, North Africa, and Asia.

ILLUSTRATIONS: Figure 112. Conard, *How to Know the Mosses* (ed. 2), fig. 220. Grout, *Mosses with Hand-Lens and Microscope*, fig. 171. Jennings, *Mosses of Western Pennsylvania* (ed. 2), Pl. 41. Welch, *Mosses of Indiana*, fig. 163.

Upper Peninsula: Chippewa Co., *Steere*. Gogebic Co., *Golley*. Houghton Co., *Parmelee*. Ontonagon Co., *Nichols* & *Steere*.
Lower Peninsula: Alpena Co., *Robinson* & *Wells*. Cheboygan Co., *Nichols*. Ingham Co., *Darlington*. Jackson Co., *Marshall*. Leelanau Co., *Darlington*. Macomb Co., *Cooley*. Oakland Co., *Darlington*. Washtenaw Co., *Kauffman*.

2. *Amblystegium juratzkanum* Schimp. — Similar to *A. serpens*; leaves wide-spreading both wet and dry, ± serrulate; costa extending beyond the leaf middle; cells up to about 6:1, basal marginal cells short-oblong to rectangular; spores in spring. — On various substrata in moist or swampy places. Nearly throughout North America; Europe; Caucasus.

ILLUSTRATIONS: Jennings, *Mosses of Western Pennsylvania* (ed. 2), Pl. 41. Welch, *Mosses of Indiana*, fig. 164.

Upper Peninsula: Chippewa Co., *Thorpe*. Mackinac Co., *Kauffman*. Ontonagon Co., *Darlington*.

3. *Amblystegium varium* (Hedw.) Lindb. — Similar to *A. serpens;* costa slender but long, ending in the acumen or subpercurrent; spores in spring. — On various substrata in wet, shady places. Almost throughout North America; Europe, Madeira, and Asia.

ILLUSTRATIONS: Conard, *How to Know the Mosses* (ed. 2), fig. 221. Grout, *Mosses with Hand-Lens and Microscope*, Pl. 78. Jennings, *Mosses of Western Pennsylvania* (ed. 2), Pl. 41. Welch, *Mosses of Indiana*, fig. 165.

Upper Peninsula: Keweenaw Co., *Cooper.* Ontonagon Co., *Darlington.*
Lower Peninsula: Alpena Co., *Robinson & Wells.* Cheboygan Co., *Nichols.* Clinton Co., *Parmelee.* Eaton, Ingham, and Leelanau counties, *Darlington.* Washtenaw Co., *Kauffman.*

4. *Amblystegium compactum* (C.M.) Aust. — Plants very small, usually in dense, yellow-green tufts or mats; stems radiculose; leaves ovate-lanceolate, gradually acuminate, denticulate all around, the teeth at the base larger and often recurved or double because of the projection of adjacent cells; costa percurrent; cells linear-rhomboidal; capsules nearly or quite erect and symmetric; spores in spring. — On decayed wood, bases of trees, and limestone in moist, shaded places. Across Canada and the northern United States; Europe.

ILLUSTRATIONS: Conard, *How to Know the Mosses* (ed. 2), fig. 222a–c. Grout, *Mosses with Hand-Lens and Microscope*, fig. 172.

Upper Peninsula: Alger and Marquette counties, *Nichols.*
Lower Peninsula: Cheboygan Co., *Nichols.* Leelanau Co., *Darlington.* Van Buren Co., *Kauffman.*

3. *Hygroamblystegium* Loeske

Aquatic plants closely related to *Amblystegium;* rather rigid, dark-green, with costa very strong, percurrent or excurrent; leaf cells with thicker walls, the alar often colored and somewhat inflated.

1. Costa excurrent in some or all leaves 2
1. Costa percurrent or nearly so 3
  2. Leaves with excurrent costa lanceolate, other leaves
    often broader . . . . . . 3a. *H. tenax* var. *spinifolium*
  2. Leaves broadly oblong or triangular-
    ovate . . . . 2. *H. noterophilum*
3. Leaves acuminate . . . . 3. *H. tenax*
3. Leaves acute or subobtuse 4
  4. Leaves oblong-lanceolate to ovate, rather
    bluntly pointed; costa strong, scarcely
    tapered to the apex . . . . 1. *H. fluviatile*
  4. Leaves ovate, acute; costa somewhat
    narrower . . . . 4. *H. orthocladum*

1. *Hygroamblystegium fluviatile* (Hedw.) Loeske — Dark-green or blackish plants with long, parallel branches; leaves oblong-lanceolate to oblong-ovate, gradually tapered to a blunt apex, entire; costa very strong, percurrent and merging into the apex, scarcely tapered; upper cells about 4–6:1; basal cells thick-walled and colored; spores in spring. — On wet rocks or soil in or beside streams. Newfoundland to Minnesota and North Carolina; Europe, North Africa, Asia, Central and South America.

ILLUSTRATIONS: Figure 113. Grout, *Mosses with Hand-Lens and Microscope*, Pl. 79 (as *Amblystegium*). Jennings, *Mosses of Western Pennsylvania* (ed. 2), Pl. 43. Welch. *Mosses of Indiana*, fig. 172.

Upper Peninsula: Keweenaw Co., *Steere*. Marquette Co, *Nichols*. Ontonagon Co., *Darlington & Bessey.*
Lower Peninsula: Hillsdale Co., *Wheeler.*

2. *Hygroamblystegium noterophilum* (Sull. & Lesq.) Warnst. — Similar to *H. fluviatile;* larger and coarser, harsh and rigid; leaves of submerged plants oblong-lanceolate (relatively narrower and longer than those of emergent plants), with a very thick, excurrent costa; spores in June. — In springs and brooks (especially in calcareous areas). Ontario and northern United States.

ILLUSTRATIONS: Conard, *How to Know the Mosses* (ed. 2), fig. 206a–b. Grout, *Moss Flora of North America*, vol. 3, Pl. 17B.

Upper Peninsula: Alger Co., *Steere.*
Lower Peninsula: Charlevoix Co., *Hill.*

3. *Hygroamblystegium tenax* (Hedw.) Jenn. — Plants medium-sized to relatively large, dark olive- to black-green; leaves broadly ovate, rather long-acuminate and acute; costa strong, percurrent or merging into the apex, usually tapered; median cells up to 6:1, 1 or more rows of basal cells somewhat enlarged; spores in spring. — On wet stones, soil, twigs, etc. in or beside brooks, often submerged. Widespread in North America; Europe, North Africa, Asia, South America.

ILLUSTRATIONS: Conard, *How to Know the Mosses* (ed. 2), fig. 219a–d. (as *H. irriguum*). Grout, *Mosses with Hand-Lens and Microscope*, fig. 174 (as *Amblystegium irriguum*). Jennings, *Mosses of Western Pennsylvania* (ed. 2), Pl. 43. Welch, *Mosses of Indiana*, fig. 169.

Upper Peninsula: Keweenaw Co., Marquette Co., *Nichols.* Ontonagon Co., *Darlington & Bessey.*
Lower Peninsula: Cheboygan Co., *Nichols.* Ionia Co., *Schnooberger.*

3a. *Hygroamblystegium tenax* var. *spinifolium* (Schimp.) Jenn. — Stem leaves lanceolate, with a thick, excurrent costa in all leaves or only in submerged leaves. — On stones, etc., in or beside streams. Eastern North America; Europe.

Upper Peninsula: Alger Co., *Nichols*. Keweenaw Co., *Steere*. Mackinac Co., *Nichols*.

4. *Hygroamblystegium orthocladum* (P.-Beauv.) Loeske — Plants similar to *H. tenax* but smaller; leaves relatively shorter and broader, cordate-ovate, shortly and broadly acuminate or gradually narrowed to a subobtuse apex. — On stones in brooks. Eastern North America; Arizona.

Upper Peninsula: Keweenaw Co., *Steere*. Marquette Co., *Nichols*. Ontona-Co., *Darlington & Bessey*.
Lower Peninsula: Cheboygan Co., *Nichols*. Ingham and Leelanau counties, *Darlington*.

4. *Cratoneuron* (Sull.) Spruce

Plants in loose or dense, green or yellowish tufts or mats, irregularly to pinnately branched, erect or prostrate; paraphyllia present, usually numerous; leaves generally plicate and ± secund and decurrent, triangular-lanceolate, short- or long-acuminate; margins ± toothed; costa single, strong, sometimes percurrent or excurrent; cells prosenchymatous, smooth (in American species), the alar cells differentiated, often inflated and forming auricles; setae elongate; capsules cylindric, strongly inclined, ± curved; peristome double and complete.

Leaves not plicate, erect or slightly secund; paraphyllia
few and inconspicuous . . . . 1. *C. filicinum*
Leaves plicate, strongly falcate-secund;
paraphyllia abundant . . . . 2. *C. commutatum*

1. *Cratoneuron filicinum* (Hedw.) Roth — Plants prostrate to erect, in loose mats, regularly pinnate to irregularly branched; paraphyllia few, not very noticeable; stem leaves cordate-triangular, slenderly acuminate, not plicate, erect-spreading to slightly secund; branch leaves narrower and somewhat more secund; costa subpercurrent to slightly excurrent; cells 3–6:1, alar cells abruptly inflated; spores in spring. — On soil, rocks, and logs in seepage or overflow of springs or in ditches, especially in calcareous habitats. Across North America; Europe, North Africa, and Asia.

[145]

ILLUSTRATIONS: Figure 114. Conard, *How to Know the Mosses* (ed. 2), fig. 197a–c. Grout, *Mosses with Hand-Lens and Microscope*, Pl. 75. Welch, *Mosses of Indiana*, fig. 174.

Upper Peninsula: Alger Co., *Wheeler.* Keweenaw Co., *Hermann.* Menominee Co., *Hill.*

Lower Peninsula: Cheboygan Co., *Nichols.* Eaton and Emmet counties, *Wheeler.* Leelanau Co., *Darlington.* Presque Isle and Washtenaw counties, *Steere.*

2. *Cratoneuron commutatum* (Hedw.) Roth — Stems suberect in dense tufts, yellowish, often encrusted with lime, densely paraphyllose and radiculose, typically pinnately branched; stem leaves broadly cordate-triangular, abruptly narrowed to a long, channeled acumen, falcate-secund, deeply plicate; costa ending in the acumen; upper cells linear or oblong-linear; alar cells ± inflated in decurrent auricles; spores in spring. — In seepage of springs and other very wet places; a calciphile. Canada and northern United States, south to Vermont, the Great Lakes, and Colorado; Europe, North Africa, and Asia.

ILLUSTRATIONS: Conard, *How to Know the Mosses* (ed. 2), fig. 197d. Grout, *Mosses with Hand-Lens and Microscope*, fig. 164.

Upper Peninsula: Alger and Mackinac counties, *Steere.*
Lower Peninsula: Leelanau Co., *Darlington.* Otsego Co., *Schnooberger.*

5. *Campylium* (Sull.) Mitt.

Plants small or rarely rather robust, in soft, loose, yellowish mats; stems creeping, freely branched; leaves lanceolate or ovate and acuminate, typically wide-spreading or squarrose, often striate; costa various, single or double, sometimes very short; cells smooth, linear, sometimes ± differentiated at the basal angles; setae elongate; capsules strongly inclined, curved; peristome double and complete.

1. Costa single, well developed 2
1. Costa short and double or lacking 3
   2. Leaves not crowded, erect-spreading . . . 1. *C. polygamum*
   2. Leaves ± crowded, wide-spreading or
     squarrose . . . . 3. *C. chrysophyllum*
3. Small plants (stems usually less than 1 cm. long; leaves
   only about 0.8 mm. long); leaves not striate . 2. *C. hispidulum*
3. Rather large plants (stems up to 10 cm. long; leaves
   1.5–3 mm. long); leaves finely striate when dry . 4. *C. stellatum*

1. *Campylium polygamum* (BSG) C. Jens. — Plants green or yellowish, moderate in size; leaves erect-spreading, rather distant, broadly lanceolate, gradually long-acuminate, about 2–2.5 mm. long, entire; costa single, ending at midleaf or beyond; cells about 8–12:1;

alar cells inflated in distinct auricles; spores in spring or early summer. — On soil in meadows, swamps, or wooded bogs. Across Canada and the northern United States to Virginia; Europe and Asia.

ILLUSTRATIONS: Conard, *How to Know the Mosses* (ed. 2), fig. 227. Grout, *Mosses with Hand-Lens and Microscope*, fig. 170. Jennings, *Mosses of Western Pennsylvania* (ed. 2), Pl. 46. Welch, *Mosses of Indiana*, fig. 178.

Upper Peninsula: Chippewa Co., *Steere.* Keweenaw Co., *Cooper.*
Lower Peninsula: Cheboygan Co., *Nichols.*

2. *Campylium hispidulum* (Brid.) Mitt. — Slender, bright-green or yellowish plants in loose mats; leaves widely spreading to squarrose, triangular-cordate and abruptly acuminate, ± serrulate all around; costa nearly or quite lacking; median cells 3–6:1; alar cells subquadrate, numerous; spores in late spring. — On soil, decaying wood, or bases of trees, sometimes on rocks in shaded places. Widespread in eastern North America, westward across Canada to British Columbia; tropical America; Europe and Asia.

ILLUSTRATIONS: Conard, *How to Know the Mosses* (ed. 2), fig. 256. Grout, *Mosses with Hand-Lens and Microscope*, fig. 168. Jennings, *Mosses of Western Pennsylvania* (ed. 2), Pl. 45. Welch, *Mosses of Indiana*, fig. 175.

Upper Peninsula: Chippewa Co., *Steere.* Gogebic Co., *Golley.* Keweenaw Co., *Gilly & Parmelee.* Marquette Co., *Nichols.* Ontonagon Co., *Nichols & Steere.*
Lower Peninsula: Alpena Co., *Wheeler.* Cass Co., *Darlington.* Cheboygan Co., *Nichols.* Clinton Co., *Parmelee.* Emmet Co., *Gilly & Parmelee.* Ingham Co., *Wheeler.* Leelanau Co., *Darlington.* Macomb Co., *Cooley.* Midland Co., *Gilly & Parmelee.* Oscoda Co., *Parmelee.* Presque Isle Co., *Steere.* St. Clair Co., *Gilly & Parmelee.* Washtenaw Co., *Kauffman.*

2a. *Campylium hispidulum* var. *sommerfeltii* (Myr.) Lindb. — Very slender plants; leaves narrower, the acumen about twice as long as the base, cells longer, and differentiated alar cells fewer. — On various substrata in moist, shady places. Canada and northern United States, south to Colorado; Europe and Asia.

Lower Peninsula: Cheboygan Co., *Nichols.*

3. *Campylium chrysophyllum* (Brid.) Bryhn — Rather slender or medium-sized plants; leaves squarrose from a somewhat clasping base (occasionally somewhat secund), ovate or ovate-lanceolate, rather abruptly narrowed to a long, channeled acumen, entire or slightly denticulate at base; costa single, ending at or above the leaf middle; cells 4–6:1, a small group of alar cells subquadrate; spores in spring or early summer. — On soil, humus, logs, etc., in moist, shady places. Eastern North America; southwestern United States to Guatemala; Europe and Asia.

[147]

ILLUSTRATIONS: Figure 115. Conard, *How to Know the Mosses* (ed. 2), fig. 216. Grout, *Mosses with Hand-Lens and Microscope*, fig. 169. Jennings, *Mosses of Western Pennsylvania* (ed. 2), Pl. 45. Welch, *Mosses of Indiana*, fig. 177.

Upper Peninsula: Chippewa Co., *Steere.* Gogebic Co., *Bessey.* Keweenaw Co., *Hermann.* Marquette Co., *Nichols.* Ontonagon Co., *Darlington & Bessey.*
Lower Peninsula: Alpena Co., *Wheeler.* Cheboygan and Emmet counties, *Nichols.* Leelanau Co., *Darlington.* Presque Isle Co., *Steere.* Washtenaw Co., *Steere.*

4. **Campylium stellatum** (Hedw.) C. Jens. — Plants rather robust, in loose, soft, shiny, yellow-green or golden tufts or mats; leaves about 1.5–3 mm. long, gradually or rather abruptly long-acuminate from an ovate or cordate base, wide-spreading or squarrose, striate when dry; costa none or occasionally present and weak or forking; cells about 6–12:1; alar cells hyaline and inflated; spores in spring. — On wet soil in swampy places, often in the open. Greenland; across the continent, south to Pennsylvania and Colorado; Europe and Asia.

ILLUSTRATIONS: Conard, *How to Know the Mosses* (ed. 2), fig. 266. Grout, *Mosses with Hand-Lens and Microscope*, Pl. 77. Jennings, *Mosses of Western Pennsylvania* (ed. 2), Pl. 46. Welch, *Mosses of Indiana*, fig. 176.

Upper Peninsula: Keweenaw Co., *Cooper.*
Lower Peninsula: Alpena Co., *Wheeler.* Cheboygan and Emmet counties, *Nichols.* Presque Isle Co., *Steere.* Van Buren Co., *Kauffman.* Washtenaw Co., *Steere.*

6. *Hygrohypnum* Lindb.

Plants typically on rocks in or beside streams, often ± submerged, creeping, irregularly branched; leaves often somewhat secund, concave, sometimes cucullate at apex, mostly broad, obtuse or rounded, sometimes apiculate or short-pointed, only rarely acute or acuminate, ± decurrent, usually entire; costa short and double or longer and forked, but often variable with some or many leaves unicostate; cells smooth, linear, the alar cells ± differentiated, sometimes conspicuously inflated; setae elongate; capsules ovoid or subcylindric, asymmetric, inclined or horizontal; peristome double, perfect (or the cilia rudimentary or lacking); annulus present.

1. Outer stem cells large, thin-walled,
   and hyaline . . . . 1. *H. ochraceum*
1. Outer stem cells not much differentiated 2
   2. Costa strong, single or forking, reaching about
      the leaf middle, rarely beyond . . . . . 2. *H. luridum*
   2. Costa none or short and double or forking 3
3. Leaves lanceolate to ovate-lanceolate;
   alar cells inflated . . . . 3. *H. eugyrium*
3. Leaves broad; alar cells not inflated 4

[148]

4. Plants soft; leaves broadly ovate, not secund,
   erect-spreading or spreading . . . . . . 4. *H. molle*
4. Plants stiff and harsh; leaves ovate to suborbicular,
   wide-spreading, often ± secund . . . . 5. *H. dilatatum*

1. *Hygrohypnum ochraceum* (Turn.) Loeske — Plants green or yellow-green; outer stem leaves enlarged, forming a hyaline sheath; leaves falcate-secund, oblong-lanceolate or oblong-ovate, gradually acuminate, blunt, denticulate at the apex; costa variable, single or double, often reaching the leaf middle or beyond; alar cells abruptly inflated in conspicuous, decurrent auricles; spores in summer. — On stones and ledges in or near brooks. Canada and the United States south to West Virginia, Colorado, and Utah; Europe and Asia.

ILLUSTRATIONS: Conard, *How to Know the Mosses* (ed. 2), fig. 270. Grout, *Mosses with Hand-Lens and Microscope*, Pl. 80. Jennings, *Mosses of Western Pennsylvania* (ed. 2), Pl. 45.

Upper Peninsula: Alger and Keweenaw counties, *Steere*. Marquette Co., *Nichols*. Ontonagon Co., *Nichols & Steere*.

2. *Hygrohypnum luridum* (Hedw.) Jenn. — Plants dark or yellow-green; leaves crowded, imbricate or more often secund, oblong-ovate, acuminate, blunt or acute, very concave with strongly incurved upper margins, entire; costa variable, single and reaching the leaf middle or beyond or forking, sometimes short, sometimes long; differentiated alar cells few, slightly inflated; spores in summer. — On wet rocks and sometimes logs in or along streams. Canada and northern United States, south to New Jersey and Colorado; Europe and Asia.

ILLUSTRATIONS: Figure 116. Conard, *How to Know the Mosses* (ed. 2), fig. 271c, 273e (as *H. palustre*). Grout, *Moss Flora of North America*, vol. 3, Pl. 23 (as *H. palustre*). Welch, *Mosses of Indiana*, fig. 179 (as *H. palustre*).

Upper Peninsula: Alger Co., *Nichols*. Keweenaw Co., *Steere*. Marquette Co., *Nichols*. Ontonagon Co., *Nichols & Steere*.
Lower Peninsula: Cheboygan Co., *Nichols*.

2a. *Hygrohypnum luridum* var. *subsphaericarpon* (Schleich.) C. Jens. — Plants robust with large, falcate-secund leaves, tapered and subtubulose above; costa thick, ending at ¾ the length of the leaf or nearly percurrent; capsules short and thick, nearly as broad as long. — On wet rocks in or near streams. Connecticut, Michigan, and British Columbia; Europe and Asia.

Upper Peninsula: Keweenaw Co. (Isle Royale), *Holt*.

3. *Hygrohypnum eugyrium* (BSG) Loeske — Plants bright-green or yellowish; leaves loosely imbricate when dry, erect-spreading

[149]

when moist, usually ± secund at the ends of stems and branches, very concave with margins infolded above, giving the stems and branches a turgid appearance, oblong-ovate to oblong-lanceolate, acute, often denticulate at the apex; costa very short and double or lacking; alar cells abruptly inflated; spores in spring. — On rocks in streams. Newfoundland to Alaska, south to North Carolina and Colorado; Europe and Asia.

ILLUSTRATIONS: Conard, *How to Know the Mosses* (ed. 2), fig. 273a–d. Grout, *Mosses with Hand-Lens and Microscope*, fig. 179.

Upper Peninsula: Keweenaw Co., *Steere.*

4. *Hygrohypnum molle* (Hedw.) Loeske — Plants soft, olive-green or brownish; leaves spreading or erect-spreading (not secund), broadly ovate, rounded-obtuse, entire or serrulate above; costa bi- to trifid, 1 branch sometimes reaching the leaf middle or beyond; alar cells small, subquadrate; spores in summer. — On rocks in streams. Pacific North America; Michigan; Europe and Asia.

ILLUSTRATION: Grout, *Moss Flora of North America*, vol. 3, Pl. 22C.

Upper Peninsula: Keweenaw Co., *Steere.* Ontonagon Co., *Nichols & Steere.*

5. *Hygrohypnum dilatatum* (Wils.) Loeske — Very similar to *H. molle;* plants stiff and harsh to the touch when dry, bright-green above, brown or blackish below; leaves wide-spreading, often ± secund, rounded-ovate (little longer than broad), rounded-obtuse and sometimes apiculate, often serrulate above. — On stones in brooks. Across the continent, south to West Virginia and Arizona; Europe and Asia.

ILLUSTRATIONS: Conard, *How to Know the Mosses* (ed. 2), fig. 274. Grout, *Mosses with Hand-Lens and Microscope*, fig. 180.

Upper Peninsula: Keweenaw Co., *Povah.* Marquette Co., *Nichols.*

7. *Scorpidium* (Schimp.) Limpr.

Very robust, turgid plants, dark- or brownish-green to blackish, sometimes reddish or golden; stems sparingly branched, prostrate; branches curved at the ends; leaves crowded, erect, somewhat secund at the tips, rugose when dry, broadly oblong-ovate, abruptly narrowed to an obtuse or apiculate apex, entire; costa lacking or faint, short, and double; cells linear, a few at the extreme angles ± inflated; setae very long; capsules strongly curved and inclined; annulus large; peristome double and perfect; spores in late autumn (rarely fruiting). — A small genus with a single species in North

America, *S. scorpioides* (Hedw.) Limpr. — In shallow water, often submerged, at margins of pools in fens or swales; a calciphile; Greenland; across the continent, south to Ohio and Montana; Europe; eastern Siberia; South America.

ILLUSTRATIONS: Figure 117. Conard, *How to Know the Mosses* (ed. 2), fig. 267. Grout, *Moss Flora of North America*, vol. 3, Pl. 27C.

Upper Peninsula: Alger Co., *Steere.* Keweenaw Co., *Allen & Stuntz.*
Lower Peninsula: Cheboygan Co., *Nichols.* Macomb Co., *Cooley.* Washtenaw Co., *Steere.*

8. *Calliergon* (Sull.) Kindb.

Rather large plants in loose tufts in swampy places; stems erect or ascending, irregularly to pinnately branched, with few radicles; leaves spreading to appressed, usually concave, ovate to suborbicular, rounded-obtuse and often cucullate at apex, sometimes apiculate; margins plane or erect, entire; costa thin and faint to strong and subpercurrent; cells linear, the alar cells sometimes abruptly inflated in conspicuous auricles; setae elongate; capsules ± curved and strongly inclined; peristome double, well developed.

1. Alar cells not much differentiated, thick-walled
   and ± colored; costa thin, often faint 2
1. Alar cells hyaline, thin-walled, ± inflated;
   costa strong 3
    2. Plants rigid, dark-colored; leaves
       rounded at apex . . . . 5. *C. trifarium*
    2. Plants soft, light- or yellow-green; leaves obtuse,
       mostly minutely apiculate . . . . . . 6. *C. turgescens*
3. Leaves narrow, oblong-lingulate . . . . 4. *C. stramineum*
3. Leaves nearly as broad as long 4
    4. Costa extending 2/3–3/4 the length
       of the leaf . . . . 3. *C. richardsonii*
    4. Costa ending near the leaf apex 5
5. Plants autoicous, irregularly branched . . . 1. *C. cordifolium*
5. Plants dioicous, pinnately branched . . . . 2. *C. giganteum*

1. *Calliergon cordifolium* (Hedw.) Kindb. — Often tall plants (up to 20 cm. high) in loose, light- or yellow-green tufts, irregularly branched; leaves ± spreading, crowded and appressed at the branch tips, oblong-ovate to cordate-ovate, rounded at apex; costa ending near the apex; alar cells enlarged and inflated in well-marked hyaline groups; autoicous; spores in late spring. — In or at margins of temporary pools and other swampy habitats. Greenland; across the continent, south to New Jersey and Illinois; Europe, Asia, and New Zealand.

[151]

ILLUSTRATIONS: Conard, *How to Know the Mosses* (ed. 2), fig. 205. Grout, *Mosses with Hand-Lens and Microscope*, Pl. 76. Jennings, *Mosses of Western Pennsylvania* (ed. 2), Pl. 44. Welch, *Mosses of Indiana*, fig. 180.

Upper Peninsula: Keweenaw Co., *Cooper.* Marquette Co., *Nichols.* Ontonagon Co., *Darlington.*
Lower Peninsula: Cheboygan Co., *Nichols.* Washtenaw Co., *Steere.*

2. *Calliergon giganteum* (Schimp.) Kindb. — Very similar to *C. cordifolium;* branching, distantly pinnate; alar cells more abruptly differentiated; dioicous; spores in late spring. — Emergent from shallow, usually temporary pools, in ditches, depressions in woods, etc. Greenland; across the continent, south to Pennsylvania; Europe and Asia.

ILLUSTRATION: Grout, *Moss Flora of North America*, vol. 3, Pl. 19A.

Upper Peninsula: Houghton Co., *Parmelee.* Luce Co., *Nichols.*
Lower Peninsula: Cheboygan Co., *Nichols.* Macomb Co., *Cooley.* Presque Isle and Washtenaw counties, *Steere.*

3. *Calliergon richardsonii* (Mitt.) Kindb. — Plants irregularly branched and similar to *C. cordifolium* but autoicous and having a shorter costa (only about 2/3–3/4 the leaf length). — On boggy soil in wet depressions or in shallow pools in swamps and cedar bogs. Greenland; northern North America south to Vermont and Michigan; Europe and Asia.

ILLUSTRATION: Grout, *Moss Flora of North America*, vol. 3, Pl. 17A (fig 10–12).

Upper Peninsula: Keweenaw Co., *Cooper.* Marquette Co., *Nichols.*

4. *Calliergon stramineum* (Brid.) Kindb. — Rather slender, light-green or straw-colored, sparsely branched, usually cuspidate at the stem or branch tips; leaves crowded, appressed and somewhat plicate when dry, oblong-lingulate, rounded-obtuse; costa 3/4–5/6 the leaf length; alar cells hyaline, abruptly inflated; spores in late spring. — In damp depressions or shallow pools at the edge of *Sphagnum* bogs or in boggy woods. Greenland; across the continent, south to Vermont, Michigan, and Wyoming; Europe and Asia.

ILLUSTRATIONS: Figure 118. Grout, *Moss Flora of North America*, vol. 3, Pl. 20.

Upper Peninsula: Alger and Keweenaw counties, *Steere.* Luce Co., *Nichols.*
Lower Peninsula: Cheboygan and Emmet counties, *Steere.*

5. *Calliergon trifarium* (Web. & Mohr) Kindb. — Rather rigid plants, brown or red-brown except at the growing tips, sparsely

branched, turgid; leaves imbricate wet or dry, often finely striate when dry, rounded-ovate, very concave, very obtuse; costa very thin but extending beyond the leaf middle; alar cells scarcely differentiated; spores in early summer. — Emergent from shallow, stagnant pools or among sedges around beach pools and in fens; a calciphile. Greenland; across Canada, south to Connecticut and Ohio; Europe and Asia.

ILLUSTRATION: Grout, *Moss Flora of North America*, vol. 3, Pl. 21.

Upper Peninsula: Mackinac Co., *Steere.*
Lower Peninsula: Emmet Co., *Nichols.* Presque Isle Co., *Steere.*

6. *Calliergon turgescens* (T. Jens.) Kindb. — Plants soft, light- or yellow-green, sometimes rather large, turgid, sparsely branched; leaves loosely or densely imbricate, oblong-ovate, very concave, cucullate and obtuse at the apex, often with a minute, recurved apiculus; costa short and thin, about 1/3 the leaf length, sometimes shorter and double or forked; alar cells few, small, rounded-quadrate; sporophytes not known from North America. — In very wet places in calcareous bogs (or fens), often emergent from shallow water. Greenland; across the continent, south to the Great Lakes; Europe and Asia.

ILLUSTRATION: Grout, *Moss Flora of North America*, vol. 3, Pl. 26B.

Lower Peninsula: Mackinac Co. (Bois Blanc I.), *Steere.*

9. *Calliergonella* Loeske

Plants in rather loose tufts, green or yellow-green, shiny, freely and sometimes pinnately branched; leaves oblong-ovate, obtuse or rounded with margins often inrolled at or near the apex; costa none or short and double; cells linear, the alar cells enlarged, sometimes inflated; setae elongate; capsules curved and inclined; peristome double, perfect.

Alar cells abruptly inflated, thin-walled, and usually
    hyaline in decurrent auricles; stems yellow . . . . 1. *C. cuspidata*
Alar cells somewhat enlarged but not thin-walled or
    in auricles; stems red . . . . 2. *C. schreberi*

1. *Calliergonella cuspidata* (Hedw.) Loeske — Light-green or yellowish plants with tips of stems and branches cuspidate because of crowded, appressed leaves; alar cells hyaline, thin-walled, and inflated in well-marked auricles; spores in late spring or early summer. — On humus in bogs, fens, and other swampy places. Across the continent, south to Tennessee; Jamaica; Europe, Africa, Asia, Australia, and New Zealand.

[153]

ILLUSTRATIONS: Figure 119. Conard, *How to Know the Mosses* (ed. 2), fig. 268. Grout, *Mosses with Hand-Lens and Microscope*, fig. 167 (as *Calliergon*).

Upper Peninsula: Alger Co., *Wheeler*. Chippewa Co., *Steere*. Keweenaw Co., *Gilly* & *Parmelee*.

Lower Peninsula: Cheboygan Co., *Nichols*. Clinton Co., *Parmelee*. Eaton Co., *Wheeler*. Oscoda Co., *Parmelee*.

2. *Calliergonella schreberi* (Brid.) Grout — Plants green or yellowish; tips of stems and branches not particularly cuspidate; alar cells somewhat enlarged and inflated, colored, not thin-walled, not in conspicuous auricles; spores in autumn. — On humus, soil, or stumps, etc., typically in dry coniferous forests. Greenland; across the continent, south to Oregon, Arkansas, and North Carolina; Europe, Asia, Central and South America.

ILLUSTRATIONS: Conard, *How to Know the Mosses* (ed. 2), fig. 269. Grout, *Mosses with Hand-Lens and Microscope*, fig. 165 (as *Calliergon*). Jennings, *Mosses of Western Pennsylvania* (ed. 2), Pl. 48 (as *Hypnum*). Welch, *Mosses of Indiana*, fig. 181.

Upper Peninsula: Alger Co., *Wheeler*. Chippewa Co., *Thorpe*. Houghton and Keweenaw counties, *Parmelee*. Marquette Co., *Nichols*. Ontonagon Co., *Nichols* & *Steere*.

Lower Peninsula: Alpena Co., *Robinson* & *Wells*. Cheboygan Co., *Nichols*. Emmet Co., *Gilly* & *Parmelee*. Ingham and Leelanau counties, *Darlington*. Oscoda Co., *Parmelee*. Washtenaw Co., *Steere*.

10. *Drepanocladus* (C.M.) Roth

Slender to robust plants in loose or dense mats or tufts mostly in wet places, sometimes aquatic; stems freely and often pinnately branched; paraphyllia none; leaves ± falcate-secund (usually strongly so), lanceolate or oblong, slenderly acuminate; costa single, well developed; cells smooth, linear, often abruptly inflated, hyaline and thin-walled in well-marked alar groups; setae elongate; capsules curved and inclined; peristome double, perfect.

1. Leaves plicate when dry . . . . 1. *D. uncinatus*
1. Leaves smooth or faintly striate when dry 2
   2. Leaves entire 3
   2. Leaves serrulate at apex, base, or both 5
3. Alar cells conspicuously inflated . . . . . 4. *D. aduncus*
3. Alar cells not inflated 4
   4. Plants yellow-green; leaves striate
     when dry . . . . 2. *D. vernicosus*
   4. Plants red or green; leaves smooth . . . . 3. *D. revolvens*
5. Alar cells rather poorly differentiated in small
   groups (never extending to the costa) . . . . 5. *D. fluitans*
5. Alar cells inflated, in large groups extending
   to the costa . . . . 6. *D. exannulatus*

1. *Drepanocladus uncinatus* (Hedw.) Warnst. — Light-green or yellowish, shiny plants in loose or dense mats; leaves strongly falcate-secund, strongly plicate when dry, denticulate toward the apex; alar cells ± enlarged and hyaline but not forming auricles; perichaetial leaves very long and sheathing; spores in spring. — On rocks, logs, humus, etc., in moist or dry coniferous woods. Arctic America to Mexico in the West, Ohio in the East; Europe, North Africa, Asia, South America.

ILLUSTRATIONS: Conard, *How to Know the Mosses* (ed. 2), fig. 210a–d. Grout, *Mosses with Hand-Lens and Microscope*, Pl. 72.

Upper Peninsula: Chippewa Co., *Thorpe.* Keweenaw Co., *Parmelee.* Marquette Co., *Nichols.* Ontonagon Co., *Darlington & Bessey.*
Lower Peninsula: Alpena Co., *Hall.* Leelanau Co., *Darlington.* Washtenaw Co., *Steere.*

2. *Drepanocladus vernicosus* (Lindb.) Warnst. — Plants yellow-green, densely tufted; leaves finely striate when dry, strongly falcate-secund, entire; alar cells not differentiated; spores in spring. — On wet soil in open, swampy places. Arctic America south to Pennsylvania and Iowa; Europe, North Africa, and Asia.

ILLUSTRATIONS: Grout, *Mosses with Hand-Lens and Microscope*, Pl. 71. Welch, *Mosses of Indiana*, fig. 182.

Upper Peninsula: Keweenaw Co., *Holt.* Luce Co., *Nichols.*
Lower Peninsula: Cheboygan Co., *Nichols.* Washtenaw Co., *Steere.*

3. *Drepanocladus revolvens* (Turn.) Warnst. — Plants green or red; leaves strongly falcate-secund, smooth, entire; alar cells not differentiated; spores in spring. — On wet soil in open, swampy places. Alaska to British Columbia and Colorado, rare in the East; Europe, Asia, and South America.

ILLUSTRATIONS: Figure 120. Grout, *Mosses with Hand-Lens and Microscope*, fig. 161, *Moss Flora of North America*, vol. 3, Pl. 28B.

Upper Peninsula: Alger Co., *Steere.* Keweenaw Co., *Holt.*
Lower Peninsula: Cheboygan and Emmet counties, *Nichols.* Leelanau Co., *Darlington.* Presque Isle Co., *Steere.*

4. *Drepanocladus aduncus* (Hedw.) Warnst. — Plants variable in size and appearance; leaves ± falcate-secund, smooth, entire; alar cells hyaline and inflated in conspicuous auricles extending nearly to the costa; spores in summer. (*D. sendtneri* is considered an ecological form with a broader costa and colored, thick-walled alar cells.) — On soil in wet, often calcareous habitats. Widely distributed in North America; Europe, North Africa, Asia, and South America.

[155]

ILLUSTRATIONS: Conard, *How to Know the Mosses* (ed. 2), fig. 202a. Grout, *Mosses with Hand-Lens and Microscope*, fig. 163b. Jennings, *Mosses of Western Pennsylvania* (ed. 2), Pl. 67. Welch, *Mosses of Indiana*, fig. 183.

Upper Peninsula: Keweenaw Co., *Allen & Stuntz.* Ontonagon Co., *Nichols & Steere.*

Lower Peninsula: Alpena Co., *Kauffman.* Cheboygan Co., *Nichols.* Clinton Co., *Parmelee.* Eaton Co., *Knobloch.* Emmet Co., *Nichols.* Leelanau Co., *Darlington.* Ottawa Co., *Kauffman.* Presque Isle Co., *Nichols.*

4a. *Drepanocladus aduncus* var. *kneiffii* (BSG) Mönk. — Leaves slightly falcate at the tips of stems and branches, otherwise straight. — On soil in wet habitats. Wide-ranging in North America; Europe, Asia, and North Africa.

ILLUSTRATIONS: Conard, *How to Know the Mosses* (ed. 2), fig. 202b. Grout, Moss Flora of North America, vol. 3, Pl. 26F. Welch, *Mosses of Indiana*, fig. 185.

Upper Peninsula: Chippewa Co., *Steere.* Keweenaw Co., *Povah.*
Lower Peninsula: Cheboygan Co., *Nichols.* Washtenaw Co., *Steere.*

4b. *Drepanocladus aduncus* var. *capillifolius* (Warnst.) Wynne — Costa ± excurrent. — Usually submerged. Widespread in North America; Europe.

Lower Peninsula: Cheboygan Co., *Wynne.*

4c. *Drepanocladus aduncus* var. *polycarpus* (Bland.) Warnst. — — Lower leaf cells oblong-hexagonal. — On wet soil. Widespread in North America; Europe, Asia, and South America.

ILLUSTRATIONS: Conard, *How to Know the Mosses* (ed. 2), fig. 202c. Grout, *Mosses with Hand-Lens and Microscope*, Pl. 74. Welch, *Mosses of Indiana*, fig. 186.

Upper Peninsula: Ontonagon Co. (Porcupine Mts.), *Nichols & Steere.*

5. *Drepanocladus fluitans* (Hedw.) Warnst. — Plants variable in all respects, usually robust, frequently floating or submerged; leaves ± falcate-secund, serrulate (especially at the apex); alar cells larger, hyaline or colored, usually not sharply delimited or forming distinct auricles; spores in early summer. — On wet soil, often submerged. Throughout Canada, Alaska, and northern United States; Europe, Asia, Africa, South America, Australia, and New Zealand.

ILLUSTRATIONS: Grout, *Mosses with Hand-Lens and Microscope*, Pl. 73. Jennings, *Mosses of Western Pennsylvania* (ed. 2), Pl. 69. Welch, *Mosses of Indiana*, fig. 188.

Upper Peninsula: Chippewa Co., *Steere.* Keweenaw Co., *Holt.* Ontonagon Co., *Nichols & Steere.*
Lower Peninsula: Cheboygan Co., *Nichols.* Washtenaw Co., *Steere.*

6. *Drepanocladus exannulatus* (BSG) Warnst. — Plants often purplish, usually pinnately branched; leaves ± falcate-secund (sometimes striolate when dry), serrulate, especially at base; alar cells abruptly inflated, forming conspicuous auricles extending to the costa. — On wet soil, sometimes submerged. Across North America, south to Pennsylvania and Colorado; Europe, North Africa, and Asia.

ILLUSTRATIONS: Conard, *How to Know the Mosses* (ed. 2), fig. 203. Grout, *Mosses with Hand-Lens and Microscope*, Pl. 74 (fig. 7, 8, 16). Jennings, *Mosses of Western Pennsylvania* (ed. 2), Pl. 44. Welch, *Mosses of Indiana*, fig. 189.

Upper Peninsula: Alger Co., *Steere.* Luce Co., *Nichols.*
Lower Peninsula: Washtenaw Co., *Steere.*

6a. *Drepanocladus exannulatus* var. *rotae* (DeNot.) Loeske — Leaves filiform-acuminate with the costa subpercurrent to long-excurrent. — Partly or entirely submerged. Across the continent, south to West Virginia, Colorado, and California; Europe and Asia.

Lower Peninsula: Cheboygan Co., *Wynne.*

BRACHYTHECIACEAE

Plants loosely or densely matted or tufted; stems prostrate to ascending, rarely erect, freely to pinnately branched; leaves erect-spreading or appressed, rarely subsecund, mostly plicate, ovate-lanceolate, mostly acuminate; costa single, well developed; cells elongate, smooth or papillose at back because of projecting ends; setae elongate, often rough; capsules inclined and asymmetric or rarely erect and symmetric; peristome double and perfect (or the cilia sometimes rudimentary or lacking); calyptra cucullate, usually naked.

1. Leaves very concave, abruptly filiform-acuminate . 4. *Cirriphyllum*
1. Leaves only moderately concave or nearly plane
   (if concave, not filiform-acuminate) 2
   2. Leaves papillose at back because of projecting
      upper cell angles . . . . 1. *Bryhnia*
   2. Leaves not papillose 3
3. Capsules erect and symmetric or nearly so; cilia
   of endostome lacking . . . . 5. *Chamberlainia*
3. Capsules asymmetric, inclined; cilia present 4

[157]

1. *Bryhnia* Kaur.

Plants rather small or medium-sized, in loose, green or yellow-green mats; stems creeping, freely branched; leaves ovate or ovate-lanceolate, acute to acuminate, decurrent; costa ending at or above the leaf middle; cells shortly oblong-rhomboidal, papillose at back because of projecting ends; dioicous; setae elongate, rough; capsules ovoid-cylindric, asymmetric and inclined; peristome double and perfect; operculum conic and sometimes short-rostrate.

Branch leaves ovate or ovate-lanceolate, acute or
   short-acuminate, twisted at the apex,
   broadly decurrent . . . . 1. *B. novae-angliae*
Branch leaves narrower, longer-acuminate,
   not twisted at the apex, slightly decurrent . . . 2. *B. graminicolor*

1. *Bryhnia novae-angliae* (Sull. & Lesq.) Grout — Plants in loose, green mats, subpinnately branched; branch leaves ovate or ovate-lanceolate, acute or short-acuminate, twisted at the apex, broadly decurrent, about 0.8–1.2 mm. long; margins plane; spores in autumn. — On moist, shaded soil, rocks, or humus. Widespread in eastern North America; Europe and Asia.

ILLUSTRATIONS: Figure 121. Conard, *How to Know the Mosses* (ed. 2), fig. 166e. Grout, *Mosses with Hand-Lens and Microscope*, fig. 152. Jennings, *Mosses of Western Pennsylvania* (ed. 2), Pl. 69. Welch, *Mosses of Indiana*, fig. 136.

Upper Peninsula: Chippewa Co., *Thorpe.* Keweenaw Co., *Allen & Stuntz.* Ontonagon Co., *Darlington & Bessey.*
Lower Peninsula: Cheboygan Co., *Nichols.* Gratiot Co., *Schnooberger.* Leelanau Co., *Darlington.* Van Buren Co., *Schnooberger.*

2. *Bryhnia graminicolor* (Brid.) Grout — Plants in loose, pale-green or yellowish mats, irregularly branched; branch leaves narrowly ovate-lanceolate, slenderly acuminate, not twisted at the apex, slightly decurrent, about 0.6–0.8 mm. long; margins reflexed below; spores in autumn. — On moist, shaded rocks or soil. Wide-ranging in eastern North America.

ILLUSTRATIONS: Conard, *How to Know the Mosses* (ed. 2), fig. 166a–c. Grout, *Mosses with Hand-Lens and Microscope*, fig. 153a. Jennings, *Mosses of Western Pennsylvania* (ed. 2), Pl. 59. Welch, *Mosses of Indiana*, fig. 137.

Lower Peninsula: Alpena Co., *Robinson & Wells*. Cheboygan Co., *Nichols*. Eaton Co., *Darlington*. Washtenaw Co., *Steere*.

2. Eurhynchium BSG

Plants rather small to moderately robust, in loose or dense mats, green or yellowish; stems creeping, freely to pinnately branched; branch leaves ovate or ovate-lanceolate, obtuse or acute, ± plicate, serrulate; costa ending at or above the leaf middle, usually in a dorsal spine; median cells oblong-linear, apical cells short and rhombic; setae elongate, sometimes rough; capsules inclined and asymmetric; peristome double, perfect; operculum long-rostrate (½–¾ the length of the urn).

1. Dark-green, rigid plants, submerged in
running water . . . . 4. *E. riparioides*
1. Plants light- or yellow-green, soft, not aquatic 2
  2. Leaves rather distant; setae rough . . . . . 1. *E. hians*
  2. Leaves crowded; setae smooth 3
3. Branches ± julaceous; leaves imbricate . . 3. *E. diversifolium*
3. Branches not julaceous; leaves
erect-spreading . . . . 2. *E. pulchellum*

1. ***Eurhynchium hians*** (Hedw.) Jaeg. & Sauerb. — Plants green or yellow-green, in rather loose mats, irregularly to subpinnately branched; leaves appearing ± complanate when dry, ovate, bluntly acute or short-acuminate, twisted at the apex, serrate throughout; costa ending near the leaf apex in a dorsal spine; setae rough; spores in late autumn. — On soil in moist, shady places. Widespread in eastern North America; Europe and Asia.

ILLUSTRATIONS: Conard, *How to Know the Mosses* (ed. 2), fig. 233a–e. Grout, *Mosses with Hand-Lens and Microscope*, fig. 154. Welch, *Mosses of Indiana*, fig. 138.

Lower Peninsula: Alpena Co., *Robinson & Wells*. Ingham Co., *Darlington*. Washtenaw Co., *Kauffman*.

2. ***Eurhynchium pulchellum*** (Hedw.) Jenn. — Rather small, green or yellowish plants; branches tapered, sometimes ± flattened when dry; leaves erect-spreading, not much crowded, ovate-lanceolate, acute or obtuse, serrate above; costa extending about 4/5 up the leaf, ending in a dorsal spine; setae smooth; spores in autumn. — On soil or humus and other substrata in rather moist, shady places. Across southern Canada, south to Tennessee and Colorado; Europe, North Africa, Asia, and South America.

ILLUSTRATIONS: Figure 122. Conard, *How to Know the Mosses* (ed. 2), fig. 233f (as *E. strigosum*). Grout, *Mosses with Hand-Lens and Microscope*, Pl. 67 (as *E. strigosum*).

Upper Peninsula: Chippewa Co., *Steere*. Gogebic Co., *Bessey*. Keweenaw and Mackinac counties, *Gilly & Parmelee*. Marquette Co., *Nichols*. Ontonagon Co., *Darlington*.

Lower Peninsula: Alpena Co., *Robinson & Wells*. Cheboygan Co., *Nichols*. Leelanau Co., *Darlington*. Washtenaw Co., *Kauffman*.

2a. **Eurhynchium pulchellum** var. **praecox** (Hedw.) Dicks. — Branches short, blunt, julaceous; leaves crowded and appressed, acute or obtuse; subquadrate alar cells few. — On soil or rocks, on shady banks. Scattered in eastern North America; Mexico; Europe, North Africa, and Asia.

ILLUSTRATIONS: Grout, *Mosses with Hand-Lens and Microscope*, Pl. 67b (fig. 1–5) (as *E. strigosum* var.). Welch, *Mosses of Indiana*, fig. 140.

Upper Peninsula: Keweenaw Co., *Cooper*.

Lower Peninsula: Leelanau Co., *Darlington*. Mecosta Co., *Schnooberger*. Oakland Co., *Darlington*. Van Buren Co., *Schnooberger*. Washtenaw Co., *Steere*.

3. **Eurhynchium diversifolium** (Schleich.) BSG — Similar to *E. pulchellum* var. *praecox*; leaves rounded-obtuse; subquadrate alar cells numerous; spores in winter. — On soil and rocks. Mountains of western North America; scattered in the East; Europe, North Africa, and Asia.

ILLUSTRATION: Grout, *Moss Flora of North America*, vol. 3, Pl. 4 (fig. 1–17).

Upper Peninsula: Gogebic Co., *Bessey*. Ontonagon Co., *Nichols & Steere*.
Lower Peninsula: Cheboygan Co., *Nichols*.

4. **Eurhynchium riparioides** (Hedw.) Rich. — Plants coarse, rather rigid and gritty when dry, dark-green, aquatic; leaves ovate, acute or subobtuse, denticulate nearly all around; costa very thick at base, extending 1/2–3/4 up the leaf, often ending in a dorsal spine; setae smooth, spores in autumn. — On wet, calcareous rocks, in or near brooks; often submerged. Widespread in North America; Europe, Africa, Asia, and Mexico to South America.

ILLUSTRATIONS: Conard, *How to Know the Mosses* (ed. 2), fig. 232. Grout, *Mosses with Hand-Lens and Microscope*, fig. 155. Welch, *Mosses of Indiana*, fig. 151.

Upper Peninsula: Alger Co., *Steere*. Gogebic Co., *Bessey*. Keweenaw Co., *Hermann*. Mackinac Co., *Ehlers*. Ontonagon Co., *Darlington & Bessey*.
Lower Peninsula: Cheboygan Co., *Steere*.

3. *Rhynchostegium* BSG

Plants similar to *Eurhynchium* but differing in the slenderly acuminate leaves which are ± complanate and not at all striate when
dry, the apical cells not shorter than the median; setae smooth; operculum long-rostrate.

Rhynchostegium serrulatum (Hedw.) Jaeg. & Sauerb. — Plants
soft, green or yellow-green, flattened; branch leaves not crowded,
ovate-lanceolate, long-acuminate, serrulate all around; costa rather
thin, usually ending near the leaf middle; spores in late summer or
early autumn. — On soil and humus, rotten wood, and tree bases
in woods. Widespread in eastern North America; tropical America;
Asia.

ILLUSTRATIONS: Figure 123. Conard, *How to Know the Mosses* (ed. 2), fig.
224 (as *Eurhynchium*).

Lower Peninsula: Cheboygan Co., *Nichols*. Clinton Co., *Parmelee*. Eaton,
Ingham, and Leelanau counties, *Darlington*. Macomb Co., *Cooley*. Ottawa and
Van Buren counties, *Kauffman*.

4. *Cirriphyllum* Grout

Rather robust, glossy plants, with creeping stems and julaceous
branches; leaves very concave and appressed, abruptly filiform-
acuminate, costate to the middle or beyond; cells linear, the alar
often quadrate; sporophyte as in *Brachythecium* (operculum usually
long-rostrate).

Cirriphyllum piliferum (Hedw.) Grout — Soft, light- or yellow-
green, glossy plants; leaves loosely imbricate, oblong-ovate, very
concave, rounded and cucullate at the apex and abruptly narrowed
to a long, filiform acumen; setae very rough; operculum about as
long as the urn, long-rostrate; spores in winter or early spring. —
On soil, humus, or tree bases in moist, shady places. Eastern North
America, from New Brunswick to Washington, D.C., and the Great
Lakes; Europe, North Africa, and Asia.

ILLUSTRATIONS: Figure 124. Grout, *Moss Flora of North America*, vol. 3,
Pl. 6 (fig. 1–23).

Upper Peninsula: Marquette Co., *Steere*. Ontonagon Co., *Conard*.
Lower Peninsula: Cheboygan and Grand Traverse counties, *Steere*.

5. *Chamberlainia* Grout

A small genus sometimes included in *Brachythecium* but differing
in the erect, cylindric capsules lacking cilia in the inner peristome.

Plants of medium size; leaves 1–1.6 mm. long, lanceolate
or ovate-lanceolate, acuminate . . . . . 1. *C. acuminata*
Plants small; leaves 0.5–1 mm. long, broader in
proportion to their length, acute or
short-acuminate . . . . 2. *C. cyrtophylla*

1. *Chamberlainia acuminata* (Hedw.) Grout — Branches terete;
leaves 1–1.6 mm. long, concave, plicate, appressed, lanceolate to
ovate-lanceolate, acuminate; cells oblong-linear; spores in autumn.
— On decaying wood, bases of trees, soil and rock in shade. Widespread in eastern North America.

ILLUSTRATIONS: Conard, *How to Know the Mosses* (ed. 2), fig. 236a–e.
Grout, *Mosses with Hand-Lens and Microscope*, Pl. 65 (as *Brachythecium*).
Jennings, *Mosses of Western Pennsylvania* (ed. 2), Pl. 64. Welch, *Mosses of
Indiana*, fig. 144.

Upper Peninsula: Marquette Co., *Hill.*
Lower Peninsula: Clinton Co., *Parmelee.* Eaton and Ingham counties, *Darlington.* Macomb Co., *Cooley.* Washtenaw Co., *Steere.*

2. *Chamberlainia cyrtophylla* (Kindb.) Grout — Leaves only
0.5–1 mm. long, ovate or ovate-lanceolate, acute or short-acuminate;
cells short and oblong-hexagonal. — On bases of trees and logs in
woods. Ontario and Michigan to North Carolina.

ILLUSTRATIONS: Figure 125. Conard, *How to Know the Mosses* (ed. 2), fig.
236f. Jennings, *Mosses of Western Pennsylvania* (ed. 2), Pl. 53.

Lower Peninsula: Cheboygan Co., *Nichols.* Huron, Ionia, and Isabella
counties, *Schnooberger.*

6. *Brachythecium* BSG

Small to moderately robust plants in loose or dense mats; stems
creeping, freely branched; leaves ovate to lanceolate, acute or acuminate, somewhat concave and usually plicate or sulcate; costa
ending above the middle to subpercurrent; cells oblong-linear to
linear, rarely oblong-hexagonal; setae elongate, often rough; capsules oblong- or ovoid-cylindric, asymmetric and inclined; operculum conic to short-rostrate; peristome double and perfect.

1. Costa percurrent or nearly so 2
1. Costa ending below the apex 4
  2. Setae rough above; leaves only
     slightly decurrent . . . . 10. *B. populeum*
  2. Setae rough throughout; leaves decurrent 3
3. Plants very slender; leaf cells short
   (3–5:1) . . . . 11. *B. reflexum*
3. Plants larger; cells 8–10:1 . . . . 12. *B. bestii*

4. Setae smooth 5
4. Setae rough above or throughout 10
5. Capsules cylindric, 3–4:1, suberect . . . . 5. *B. oxycladon*
5. Capsules ovoid- or oblong-cylindric, 2–3:1,
    strongly inclined 6
6. Stem leaves gradually tapered 7
6. Stem leaves more abruptly acuminate 8
7. Stem leaves lanceolate, serrulate above . . . 2. *B. flexicaule*
7. Stem leaves triangular-ovate, subentire . . . . . 3. *B. acutum*
8. Branches turgid; leaves entire . . . . 15. *B. turgidum*
8. Branches not turgid; leaves serrulate
    (at least above) 9
9. Branch leaves lanceolate, 1.8–2.3 mm. long . . 1. *B. salebrosum*
9. Branch leaves oblong-ovate, up to 1 mm. long . 6. *B. digastrum*
10. Setae rough above, nearly smooth below 11
10. Setae roughened throughout 12
11. Leaves serrate above, plicate . . . . . 4. *B. campestre*
11. Leaves entire or nearly so, smooth . . . . 9. *B. plumosum*
12. Alar cells not or somewhat inflated, not
    decurrent groups (especially in
    stem leaves) . . . . 8. *B. rivulare*
12. Alar cells not or somewhat inflated, not
    in conspicuous groups 13
13. Small plants with leaves falcate-secund,
    up to 1.3 mm. long . . . . 14. *B. velutinum*
13. Plants moderately robust; leaves not falcate-secund,
    1.5–2 mm. long 14
14. Plants terete-foliate; leaves erect-spreading,
    distantly serrate . . . . 7. *B. rutabulum*
14. Plants complanate-foliate; leaves spreading,
    strongly serrate . . . . 13. *B. starkei*

1. *Brachythecium salebrosum* (Web. & Mohr) BSG — Plants yellow-green, terete-foliate; leaves lanceolate, long-acuminate; margins serrate above, reflexed below; costa ending beyond the leaf middle; cells linear, about 10:1; autoicous; setae smooth; spores in autumn or early winter. — On earth, stones, logs and bases of trees, often in disturbed habitats. Widespread in eastern North America; Europe, Africa, Asia, Australia, and New Zealand.

ILLUSTRATIONS: Conard, *How to Know the Mosses* (ed. 2), fig. 243. Grout, *Mosses with Hand-Lens and Microscope*, Pl. 62. Jennings, *Mosses of Western Pennsylvania* (ed. 2), Pl. 54. Welch, *Mosses of Indiana*, fig. 146.

Upper Peninsula: Chippewa Co., *Steere*. Gogebic Co., *Golley*. Houghton Co., *Parmelee*. Ontonagon Co., *Nichols* & *Steere*.
Lower Peninsula: Alpena Co., *Robinson* & *Wells*. Cheboygan Co., *Nichols*. Eaton Co., *Darlington*. Emmet Co., *Nichols*. Leelanau and Oakland counties, *Darlington*. Washtenaw Co., *Kauffman*.

2. *Brachythecium flexicaule* Ren. & Card. — Plants yellow-green, subpinnately branched; stem leaves gradually tapered to a slender apex; branch leaves lanceolate, gradually long-acuminate;

margins serrate, ± reflexed; costa 1/2–2/3 the length of the leaf; cells linear, 12–14:1; monoicous; setae smooth. — On various substrata in woods. British Columbia, Newfoundland and southeastern Canada to Pennsylvania and the Great Lakes.

ILLUSTRATIONS: Grout, *Mosses with Hand-Lens and Microscope*, fig. 144. Jennings, *Mosses of Western Pennsylvania* (ed. 2), Pl. 55. Welch, *Mosses of Indiana*, fig. 147.

Upper Peninsula: Chippewa Co., *Steere*. Houghton Co., *Parmelee*. Lower Peninsula: Cheboygan Co., *Nichols*. Ingham and Leelanau counties, *Darlington*.

3. **Brachythecium acutum** (Mitt.) Sull. — Glossy, green or yellow-green plants; leaves often somewhat complanate, lanceolate to ovate-lanceolate, gradually and slenderly acuminate, entire or distantly serrulate, not or slightly plicate; costa about 2/3 the leaf length; autoicous (or polygamous); setae smooth; spores maturing in autumn. — On earth and logs in swampy places. Across North America, south to New Jersey, Ohio, and Colorado; Europe.

ILLUSTRATIONS: Conard, *How to Know the Mosses* (ed. 2), fig. 244d. Grout, *Mosses with Hand-Lens and Microscope*, fig. 146. Welch, *Mosses of Indiana*, fig. 148.

Lower Peninsula: Clinton Co., *Parmelee*. Ingham Co., *Darlington*. Otsego Co., *Schnooberger*. Van Buren Co., *Kauffman*. Washtenaw Co. *Steere*.

4. **Brachythecium campestre** (C.M.) BSG — Plants dark-green or yellowish and glossy; leaves equally spreading to somewhat falcate-secund, lanceolate to ovate-lanceolate, long-acuminate, serrate above, plicate, costate to the middle or beyond; autoicous; setae ± roughened above, nearly smooth below; spores in autumn. — On earth and stones in damp woods or grassy places. British Columbia; widespread in eastern North America; Europe, North Africa, and Asia.

ILLUSTRATIONS: Grout, *Moss Flora of North America*, vol. 3, Pl. 8D. Jennings, *Mosses of Western Pennsylvania* (ed. 2), Pl. 54. Welch, *Mosses of Indiana*, fig. 149.

Upper Peninsula: Chippewa Co., *Steere*. Ontonagon Co., *Nichols & Steere*. Lower Peninsula: Alpena Co., *Hall*. Van Buren Co., *Schnooberger*. Washtenaw Co., *Steere*.

5. **Brachythecium oxycladon** (Brid.) Jaeg. & Sauerb. — Plants glossy, yelllow-green; leaves ovate-lanceolate, acute or acuminate, serrulate nearly all around, plicate; costa extending beyond the leaf middle; alar cells subquadrate, dense; dioicous (or rarely autoi-

[164]

cous); setae smooth; capsules oblong-cylindric, suberect; spores in autumn or early winter. — On various substrata in shady places. Widespread in eastern North America; Colorado; Europe and South America.

ILLUSTRATIONS: Conard, *How to Know the Mosses* (ed. 2), fig. 241a–c. Grout, *Mosses with Hand-Lens and Microscope*, Pl. 63. Jennings, *Mosses of Western Pennsylvania* (ed. 2), Pl. 69. Welch, *Mosses of Indiana*, fig. 150.

Upper Peninsula: Alger Co., *Wheeler.* Chippewa Co., *Thorpe.* Gogebic Co., *Golley.* Keweenaw Co., *Hermann.* Ontonagon Co., *Darlington & Bessey.* Lower Peninsula: Alpena Co., *Robinson & Wells.* Cheboygan Co., *Nichols.* Eaton Co., *Darlington.* Ingham Co., *Parmelee.* Leelanau Co., *Darlington.* Washtenaw Co., *Kauffman.*

5a. *Brachythecium oxycladon* var. *dentatum* (Lesq. & James) Grout — Plants dirty brownish-green below, slender; leaves shorter-pointed, more strongly serrate, more loosely areolate at base. — On rocks and earth in wet places (often submerged). Ranging with the species in North America.

Upper Peninsula: Marquette Co., *Nichols.* Presque Isle Co., *Steere.*

6. *Brachythecium digastrum* C. M. & Kindb. — Olive-green plants with subjulaceous branches; branch leaves mostly imbricate when dry, erect-open when moist, ovate or oblong-ovate, acute or short-acuminate, ± twisted at the apex, biplicate, decurrent; margins serrulate, reflexed below; costa ending about 2/3–3/4 up the leaf; quadrate alar cells numerous; stem leaves triangular-ovate, longer-acuminate; autoicous; setae smooth. — On rocks. Ontario and Michigan to North Carolina.

ILLUSTRATION: Grout, *Mosses with Hand-Lens and Microscope*, fig. 145.

Lower Peninsula: Clinton Co., *Darlington.* Gratiot Co., *Schnooberger.*

7. *Brachythecium rutabulum* (Hedw.) BSG — Plants glossy, green or yellowish; branches terete- to ± complanate-foliate; branch leaves ovate-lanceolate, gradually long-acuminate, distantly serrulate, scarcely plicate, slightly decurrent; costa 2/3 the leaf length; stem leaves broadly ovate, more abruptly acuminate, more decurrent; autoicous (or rarely polygamous); setae rough; spores in winter. — On soil, rocks, logs, and tree bases in moist, shady places. Canada and northern United States, south to Tennessee, Missouri, and Montana; Europe and Asia.

ILLUSTRATIONS: Conard, *How to Know the Mosses* (ed. 2), fig. 247. Grout, *Mosses with Hand-Lens and Microscope*, Pl. 64. Jennings, *Mosses of Western Pennsylvania* (ed. 2), Pl. 55. Welch, *Mosses of Indiana*, fig. 151.

[165]

Upper Peninsula: Chippewa Co., *Thorpe.* Keweenaw Co., *Cooper.* Ontonagon Co., *Darlington.*

Lower Peninsula: Alpena Co., *Robinson & Wells.* Cheboygan Co., *Nichols.* Clinton Co., *Parmelee.* Leelanau Co., *Darlington.* Macomb Co., *Cooley.* Van Buren and Washtenaw counties, *Kauffman.*

7a. *Brachythecium rutabulum* var. *flavescens* BSG — Plants stout, straw-colored, loosely foliate, turgid; stem leaves broadly ovate, abruptly short-acuminate. — Range and habitat similar to those of the species. Europe and North Africa.

Upper Peninsula: Mackinac Co. (Mackinac I.), *Nichols.*

8. *Brachythecium rivulare* BSG — Rather robust, glossy, light-green plants with creeping primary stems and somewhat dendroid, curved-ascending secondary stems; branch leaves ovate or ovate-lanceolate, acute to short-acuminate, somewhat decurrent, sharply denticulate above, often ± plicate; stem leaves broadly ovate, decurrent, rather abruptly short-acuminate; costa often forked; alar cells abruptly inflated; dioicous; setae smooth; spores in early autumn. — On wet rocks or soil in seepage or in wet or periodically inundated places. Widespread in eastern North America; Europe, Africa, Asia, and Australia.

ILLUSTRATIONS: Conard, *How to Know the Mosses* (ed. 2), fig. 248a–d. Grout, *Mosses with Hand-Lens and Microscope*, fig. 147. Jennings, *Mosses of Western Pennsylvania* (ed. 2), Pl. 56. Welch, *Mosses of Indiana*, fig. 152.

Upper Peninsula: Chippewa Co., *Steere.* Keweenaw Co., *Hermann.* Marquette Co., *Nichols.* Ontonagon Co., *Darlington.*

Lower Peninsula: Alpena Co., *Robinson & Wells.* Cheboygan Co., *Nichols.* Eaton Co., *Parmelee.* Ingham and Leelanau counties, *Darlington.* Washtenaw Co., *Steere.*

9. *Brachythecium plumosum* (Hedw.) BSG — Brownish- or golden-green plants, usually glossy; branch leaves sometimes subsecund, lanceolate to ovate-lanceolate, long-acuminate, serrulate to serrate above, sometimes subentire, smooth or nearly so; costa about 2/3 the leaf length; stem leaves ovate, acuminate, often somewhat decurrent; autoicous; setae rough above, nearly smooth below; spores in late autumn. — On moist, shaded rocks, usually in or near streams. British Columbia; southeastern Canada and the Great Lakes to Tennessee; Europe, Africa, Asia, Mexico and New Zealand.

ILLUSTRATIONS: Conard, *How to Know the Mosses* (ed. 2), fig. 246a–d. Grout, *Mosses with Hand-Lens and Microscope*, fig. 149. Jennings, *Mosses of Western Pennsylvania* (ed. 2), Pl. 57. Welch, *Mosses of Indiana*, fig. 153 (as *B. flagellare*).

Upper Peninsula: Keweenaw Co., *Povah*. Ontonagon Co., *Nichols & Steere*. Lower Peninsula: Cheboygan Co., *Nichols*. Eaton Co., *Darlington*. Washtenaw Co., *Kauffman*.

10. *Brachythecium populeum* (Hedw.) BSG — Plants rather small, dark- or yellowish-green; branch leaves lanceolate or ovate-lanceolate, subulate-acuminate, subentire or serrulate above; costa nearly or quite percurrent; cells oblong-linear, 5–8:1, several rows at basal angles rhombic to quadrate; autoicous; setae rough above, nearly smooth below; spores in early winter. — On rocks or, less often, soil or humus in woods. British Columbia; widespread in eastern North America; Europe, Africa, and Asia.

ILLUSTRATIONS: Conard, *How to Know the Mosses* (ed. 2), fig. 246e. Grout, *Mosses with Hand-Lens and Microscope*, fig. 150.

Upper Peninsula: Chippewa Co., *Thorpe*. Marquette Co., *Nichols*. Ontonagon Co., *Darlington & Bessey*.
Lower Peninsula: Cheboygan Co., *Nichols*. Washtenaw Co., *Steere*.

11. *Brachythecium reflexum* (Starke) BSG — Plants small, in dark- or yellow-green mats, pinnately branched; branch leaves loosely imbricate when dry, lanceolate, gradually acuminate, decurrent, serrate above; costa extending into the acumen; cells oblong-hexagonal, 3–5:1; quadrate alar cells very numerous; stem leaves deltoid-ovate, long-acuminate; autoicous; setae rough throughout; spores in autumn or early winter. — On rocks, soil, logs, and bases of trees in woods. Eastern North America, south to Maryland, west to Montana; Europe and Asia.

ILLUSTRATIONS: Figure 126. Conard, *How to Know the Mosses* (ed. 2), fig. 245. Grout, *Mosses with Hand-Lens and Microscope*, fig. 148.

Upper Peninsula: Baraga Co., *Nichols & Steere*. Chippewa Co., *Thorpe*. Keweenaw Co., *Povah*. Luce and Mackinac counties, *Nichols*. Ontonagon Co., *Nichols & Steere*.
Lower Peninsula: Cheboygan Co., *Nichols*. Emmet Co., *Steere*.

12. *Brachythecium bestii* Grout — Somewhat more robust than *B. reflexum* with stem leaves ovate-lanceolate and more closely appressed; cells somewhat more than twice as long; alar cells rectangular, seldom quadrate; costa generally ending in the base of the acumen but sometimes percurrent. — On soil of banks or in rock crevices in shady places. Alaska to Oregon and Idaho; Michigan, Nova Scotia, and Newfoundland.

ILLUSTRATION: Grout, *Moss Flora of North America*, vol. 3, Pl. 12 (fig. 4a–c).

Upper Peninsula: Ontonagon Co. (Porcupine Mts.), *Nichols* & *Steere.*

13. *Brachythecium starkei* (Brid.) BSG — Plants dark- or rarely whitish-green; branch leaves ± complanate, ovate-lanceolate, acute or acuminate, often twisted at apex, serrate, somewhat decurrent; costa extending beyond the leaf middle; autoicous; setae rough; spores in winter. — On moist logs and stumps. Across the continent, south to Montana and Pennsylvania; Europe and Asia.

ILLUSTRATIONS: Grout, *Moss Flora of North America,* vol. 3, Pl. 8C (fig. 4–8). Jennings, Mosses of Western Pennsylvania (ed. 2), Pl. 56.

Upper Peninsula: Alger Co., *Wheeler.* Chippewa Co., *Steere.* Mackinac Co., *Wheeler.*
Lower Peninsula: Cheboygan Co., *Nichols.*

14. *Brachythecium velutinum* (Hedw.) BSG — Plants slender, dark- or yellow-green; branch leaves loosely spreading and ± falcate-secund, lanceolate to ovate-lanceolate, gradually long-acuminate, serrate; costa extending beyond the leaf middle, often toothed at back; autoicous; setae rough; spores in autumn to early spring. — On various substrata in woods. Across the continent, south to California in the West, New Jersey in the East; Europe, North Africa, and Asia.

ILLUSTRATIONS: Conard, *How to Know the Mosses* (ed. 2), fig. 249. Grout, *Mosses with Hand-Lens and Microscope,* Pl. 66. Jennings, *Mosses of Western Pennsylvania* (ed. 2), Pl. 56.

Upper Peninsula: Keweenaw Co., *Cooper.*
Lower Peninsula: Alpena Co., *Robinson* & *Wells.* Cheboygan Co., *Nichols.* Clinton Co., *Parmelee.* Leelanau Co., *Darlington.*

15. *Brachythecium turgidum* (Hartm.) Kindb. — Rather robust, turgid plants in glossy, whitish- to golden-green mats; leaves loosely appressed, strongly plicate, lanceolate or ovate-lanceolate, filiform-acuminate, entire or somewhat serrulate, costate to the middle; autoicous; setae smooth; spores in autumn. — On wet soil or rocks near streams, waterfalls, etc. Greenland; boreal North America, south to Quebec and Michigan; Europe and Asia.

Lower Peninsula: Mackinac Co. (Bois Blanc I.), *Steere.*

7. *Camptothecium* BSG

Plants medium-sized to large, creeping to erect or ascending, often regularly pinnate, green to yellowish or brownish; leaves lanceolate, long-acuminate, deeply plicate, costate to the middle; cells very long and narrow (up to 20:1), the alar cells differentiated

[168]

(usually quadrate and thick-walled); setae smooth or rough; capsules oblong-cylindric, ± curved; peristome double and perfect; operculum conic or conic-rostrate.

Camptothecium nitens (Hedw.) Schimp. — Plants robust, shiny-green to golden-brown, erect, usually pinnately branched, brown-tomentose; leaves long, lance-acuminate, erect, deeply plicate; margins very narrowly recurved, entire or nearly so; costa 3/4 the leaf length; setae smooth; capsules curved and strongly inclined; spores in spring. — In calcareous bogs. Across the continent, south to New Jersey and the Great Lakes; Europe and Asia.

ILLUSTRATIONS: Figure 127. Conard, *How to Know the Mosses* (ed. 2), fig. 229. Grout, *Mosses with Hand-Lens and Microscope*, fig. 157.

Upper Peninsula: Keweenaw Co., *Allen & Stuntz.*
Lower Peninsula: Cheboygan Co., *Nichols.* Eaton Co., *Darlington.* Grand Traverse, Presque Isle, and Washtenaw counties, *Steere.*

ENTODONTACEAE

Mostly shiny plants in low mats, creeping, freely branched; stems and branches julaceous or ± flattened; paraphyllia none; leaves appressed, sometimes complanate, often concave or plicate, ecostate or nearly so; cells linear, smooth, the alar cells mostly quadrate and numerous; setae elongate, smooth; capsules erect and symmetric, long-cylindric; peristome double, the teeth narrow, the segments mostly linear from a low basal membrane, cilia rudimentary or lacking .

Entodon C. M.

Glossy, yellow-green, julaceous or flattened plants; leaves crowded, appressed, oblong-ovate, obtuse or acute to acuminate, entire or slightly toothed at the apex; costa short and double or none; quadrate alar cells in numerous rows; setae long and smooth; peristome inserted below the mouth.

Branches flattened, more than 1 mm. wide   .   .   1. *E. cladorrhizans*
Branches julaceous, less than 1 mm. wide   .   .   .   2. *E. seductrix*

1. *Entodon cladorrhizans* (Hedw.) C. M. — Leaves strongly complanate, oblong-ovate, acute, entire or slightly denticulate at apex; peristome teeth smooth or nearly so above, granulose below; spores in autumn. — On various substrata, especially rotten logs and leaf litter, in woods. Widespread in eastern North America; New Mexico; Europe and Asia.

[169]

ILLUSTRATIONS: Conard, *How to Know the Mosses* (ed. 2), fig. 290. Grout, *Mosses with Hand-Lens and Microscope*, fig. 204a–d. Jennings, *Mosses of Western Pennsylvania* (ed. 2), Pl. 36. Welch, *Mosses of Indiana*, fig. 214.

Lower Peninsula: Alpena Co., *Robinson & Wells.* Clinton Co., *Parmelee.* Emmet Co., *Steere.* Ingham Co., *Parmelee.* Leelanau Co., *Darlington.* Macomb Co., *Cooley.* Washtenaw Co., *Steere.*

2. **Entodon seductrix** (Hedw.) C. M. — Julaceous plants; leaves oblong-ovate, shortly apiculate, entire or slightly serrulate at apex; costa short and double; quadrate alar cells numerous; peristome teeth smooth or nearly so; spores in autumn or winter. — On rotten logs and other substrata in rather dry, shady places. Widespread in eastern North America.

ILLUSTRATIONS: Figure 128. Conard, *How to Know the Mosses,* (ed. 2), fig. 286a–g. Jennings, *Mosses of Western Pennsylvania* (ed. 2), Pl. 37.

Upper Peninsula: Ontonagon Co., *Darlington & Bessey.*
Lower Peninsula: Alpena Co., *Robinson & Wells.* Cheboygan Co., *Steere.* Ingham Co., *Parmelee.* St. Joseph Co., *Darlington.* Washtenaw Co., *Steere.*

PLAGIOTHECIACEAE

Plants in flat, mostly shiny mats; stems creeping, irregularly branched; leaves mostly strongly complanate, ovate or ovate-lanceolate, acuminate; costa short and double or lacking; cells linear, smooth, usually not differentiated at the basal angles; setae elongate; capsules oblong-ovoid to cylindric, ± asymmetric, mostly inclined; peristome double, perfect or the cilia rarely rudimentary or lacking.

Leaves decurrent; alar cells ± differentiated . . 1. *Plagiothecium*
Leaves not decurrent; alar cells not or only
slightly differentiated . . . . 2. *Isopterygium*

1. *Plagiothecium* BSG

Plants in flat, shiny, green or yellow-green mats; paraphyllia none; leaves mostly distichous-complanate, ovate-lanceolate and acuminate, decurrent; alar cells ± differentiated; capsules suberect to curved and inclined.

1. Leaves complanate 2
1. Leaves not or slightly complanate 5
    2. Plants dingy- or yellow-green; leaves shrinking
    and scarcely overlapping when dry . . . 2. *P. sylvaticum*
    2. Plants glossy-green; leaves not shrinking when dry 3
3. Edges of leaves strongly incurved, clasping the
base of the leaf above; costa rather strong, extending
about 1/3–1/2 the leaf length . . . . . . 3. *P. ruthei*
3. Edges of leaves only slightly incurved;
costa well marked 4

[170]

4. Leaves short-acuminate, rather strongly
decurrent; capsules horizontal
or nearly so . . . . 1. *P. denticulatum*
4. Leaves slenderly acuminate, slightly decurrent;
capsules suberect . . . . 4. *P. laetum*
5. Leaves loosely imbricate . . . . 5. *P. roeseanum*
5. Leaves squarrose-spreading . . . . . . 6. *P. striatellum*

1. *Plagiothecium denticulatum* (Hedw.) BSG — Shiny-green
plants; leaves strongly complanate, scarcely altered on drying;
leaves oblong-ovate, acute or short-acuminate, entire except for a
few short teeth at the tip; capsules strongly inclined, asymmetric;
spores in summer. — On soil or humus, particularly on steep, shaded
banks, and often on tree bases, logs and rocks. Across the continent,
south to Georgia and Colorado; Europe and Asia.

ILLUSTRATIONS: Conard, *How to Know the Mosses* (ed. 2), fig. 294. Grout,
Mosses with Hand-Lens and Microscope, Pl. 84. Jennings, *Mosses of Western
Pennsylvania* (ed. 2), Pl. 53. Welch, *Mosses of Indiana*, fig. 207.

Upper Peninsula: Chippewa Co., *Steere.* Houghton Co., *Parmelee.* Keweenaw
Co., *Nichols.* Ontonagon Co., *Darlington.*
Lower Peninsula: Alpena Co., *Robinson & Wells.* Cheboygan Co., *Nichols.*
Eaton Co., *Darlington.* Ingham Co., *Wheeler.* Kalamazoo Co., *Darlington.*
Leelanau, Monroe, and Oakland counties, *Darlington.* Oscoda Co., *Parmelee.*
St. Joseph Co., *Darlington.* Washtenaw Co., *Kauffman.*

2. *Plagiothecium sylvaticum* (Brid.) BSG — Similar to *P. denticulatum;* plants yellowish, less complanate-foliate; leaves shrunken
and ± contorted when dry. — On various substrata in woods.
Across the continent, south to Georgia and Colorado; Europe and
Asia.

ILLUSTRATIONS: Figure 129. Jennings, *Mosses of Western Pennsylvania* (ed.
2), Pl. 51. Welch, *Mosses of Indiana*, fig. 208.

Upper Peninsula: Marquette Co., *Nichols.* Ontonagon Co., *Nichols & Steere.*
Lower Peninsula: Alpena Co., *Robinson & Wells.* Cheboygan Co., *Roberts.*

3. *Plagiothecium ruthei* Limpr. — Much like *P. denticulatum;*
leaves strongly incurved at the margins and clasping the leaf above.
— In wet places. Widespread in eastern North America; Europe.

Upper Peninsula: Chippewa and Ontonagon counties, *Steere.*

Lower Peninsula: Reported by Steere for Cheboygan, Grand Traverse, and
Washtenaw counties.

4. *Plagiothecium laetum* BSG — Much like *P. denticulatum* but
smaller; leaves more slenderly acuminate, less decurrent, often with
narrowly revolute margins; cilia of endostome usually lacking. —

[171]

On shaded soil, etc. Southeastern Canada to Michigan and Tennessee; New Mexico; Europe and Siberia.

ILLUSTRATION: Grout, *Moss Flora of North America*, vol. 3, Pl. 40 (fig. 4–9).

Upper Peninsula: Ontonagon Co., *Nichols & Steere*.
Lower Peninsula: Washtenaw Co., *Steere*.

5. *Plagiothecium roeseanum* (Hampe) BSG — Similar to *P. denticulatum* or *P. sylvaticum;* leaves concave, erect and loosely imbricate, not complanate, the branches thus ± julaceous. — In crevices or on ledges of cliffs or soil of banks. Across the continent and south to Georgia and Colorado; Europe and Asia.

ILLUSTRATIONS: Conard, *How to Know the Mosses* (ed. 2), fig. 292. Jennings, *Mosses of Western Pennsylvania* (ed. 2), Pl. 71. Welch, *Mosses of Indiana*, fig. 209.

Upper Peninsula: Chippewa and Keweenaw counties, *Steere*. Ontonagon Co., *Nichols & Steere*.
Lower Peninsula: Alpena Co., *Robinson & Wells*. Cheboygan Co., *Nichols*. Eaton Co., *Darlington*. Ingham Co., *Parmelee*. Kalamazoo Co., *Becker*. Leelanau and St. Joseph counties, *Darlington*. Washtenaw Co., *Steere*.

6. *Plagiothecium striatellum* (Brid.) Lindb. — Plants in dense, dark-green mats; leaves squarrose-spreading, ovate-lanceolate to ovate, slenderly acuminate, decurrent, serrulate (at least above); alar cells abruptly inflated in distinct auricles; capsules suberect to curved and inclined; spores in May and June. — On stones, humus, and rotten wood in shady places. Eastern North America, south to North Carolina; Alaska to British Columbia; Europe.

ILLUSTRATIONS: Grout, *Mosses with Hand-Lens and Microscope*, fig. 200. Jennings, *Mosses of Western Pennsylvania* (ed. 2), Pl. 51.

Upper Peninsula: Marquette Co., *Nichols*.
Lower Peninsula: Cheboygan Co., *Nichols*. Washtenaw Co., *Kauffman*.

2. Isopterygium Mitt.

Plants in light-green or yellowish, mostly glossy mats; leaves mostly ± complanate, ovate-lanceolate, acute or acuminate, not or slightly decurrent; cells linear, smooth, the alar cells scarcely differentiated.

1. Leaves serrulate nearly all around 2
1. Leaves entire or slightly serrulate at the apex 3
   2. Leaves loosely complanate and usually ± secund
      at the tips, not closely overlapping . . . 5. *I. turfaceum*
   2. Leaves clearly complanate, not secund,
      closely overlapping . . . . 1. *I. deplanatum*

3. Leaves scarcely complanate, usually secund at
   the tips of branches . . . . 4. *I. pulchellum*
3. Leaves strongly complanate, not secund 4
    4. Cortical cells small and thick-walled;
      leaves ± serrulate at apex . . . . . . 2. *I. elegans*
    4. Cortical cells large and thin-walled;
      leaves entire . . . . 3. *I. muellerianum*

1. **Isopterygium deplanatum** (Sull.) Grout — Golden-green plants
in flat mats; leaves overlapping, complanate in 2 rows, oblong-
lanceolate to ovate, acuminate, sometimes broadly acute or rather
blunt at apex, serrate above, serrulate nearly to the base; median
cells linear; apical cells shorter in broadly pointed leaves. — On
earth or stones in shade. Widespread in eastern North America;
Arizona.

ILLUSTRATIONS: Conard, *How to Know the Mosses* (ed. 2), fig. 294 (as
Plagiothecium). Grout, *Mosses with Hand-Lens and Microscope*, fig. 199 (as
Plagiothecium). Jennings, *Mosses of Western Pennsylvania* (ed. 2), Pl. 69.
Welch, *Mosses of Indiana*, fig. 210 (as *Plagiothecium*).

Upper Peninsula: Alger and Chippewa counties, *Steere.*
Lower Peninsula: Alpena Co., *Robinson & Wells.* Eaton and Leelanau
counties, *Darlington.* Washtenaw Co., *Steere.*

2. **Isopterygium elegans** (Hook.) Lindb. — Plants small, glossy-
green; cortical cells small and thick-walled; leaves complanate and
appearing 2-ranked, oblong-lanceolate or ovate-lanceolate, acuminate, entire except at the apex; spores in early spring; gemmiform
branchlets often produced in leaf axils. — On humus or in crevices
of moist, shaded rocks. Across the continent, south to Tennessee and
California; Europe.

ILLUSTRATIONS: Conard, *How to Know the Mosses* (ed. 2), fig. 298e, f.
Grout, *Mosses with Hand-Lens and Microscope*, fig. 195 (as *Plagiothecium*).
Jennings, *Mosses of Western Pennsylvania* (ed. 2), Pl. 68.

Upper Peninsula: Alger and Marquette counties, *Nichols.*
Lower Peninsula: Cheboygan and Emmet counties, *Nichols.*

3. **Isopterygium muellerianum** (Schimp.) Lindb. — Plants yellow-green; cortical cells large and thin-walled; leaves strongly complanate, lanceolate or ovate-lanceolate, rather abruptly acuminate,
entire; spores in autumn. — On moist soil and rocks, especially in
crevices, in sheltered places, such as ravines. Nova Scotia to Minnesota, south to North Carolina and Ohio; Europe and Asia.

ILLUSTRATIONS: Grout, *Mosses with Hand-Lens and Microscope*, fig. 194 (as
Plagiothecium). Welch, *Mosses of Indiana*, fig. 212.

Upper Peninsula: Keweenaw Co., *Steere.* Ontonagon Co., *Nichols & Steere.*

[173]

4. *Isopterygium pulchellum* (Hedw.) Jaeg. & Sauerb. — Very small, green or yellow-green plants with branches suberect and usually curved; leaves not complanate, often secund at branch tips, lanceolate or ovate-lanceolate, gradually acuminate, entire; spores in early summer. — On humus or in crevices of moist, shaded rock. Across the continent, south to Pennsylvania, Colorado, and Washington; Europe.

Illustrations: Figure 130. Grout, *Mosses with Hand-Lens and Microscope,* fig. 197 (as *Plagiothecium*).

Upper Peninsula: Chippewa Co. (Sugar I.), *Steere.*

5. *Isopterygium turfaceum* (Lindb.) Lindb. — Plants bright-green or yellowish, loosely matted; leaves loosely complanate and often subsecund, ovate- or oblong-lanceolate, acuminate, serrate above, serrulate nearly to the base; spores in summer. — On decaying wood and humus in moist, shady places. Nova Scotia to Manitoba, south to Pennsylvania; Europe and Asia.

Illustrations: Grout, *Mosses with Hand-Lens and Microscope,* Pl. 83. Jennings, *Mosses of Western Pennsylvania* (ed. 2), Pl. 51.

Upper Peninsula: Keweenaw Co., *Cooper.* Marquette Co., *Nichols.* Ontonagon Co., *Darlington.*
Lower Peninsula: Cheboygan Co., *Nichols.* Ingham and Leelanau counties, *Darlington.* Van Buren Co., *Kauffman.*

Sematophyllaceae

Slender to moderately robust plants in green or yellowish, often shiny mats; stems creeping or rarely ascending, freely branched, sometimes pinnate; leaves often complanate, mostly homomallous or falcate-secund, ovate or oblong-lanceolate, mostly acute; costa very short and double or lacking; cells mostly linear or rhomboidal, smooth or variously papillose, nearly always clearly differentiated at the basal angles; setae elongate, smooth or ± papillose; capsules mostly inclined and asymmetric; operculum usually rostrate; peristome double (or rarely single), perfect (or the cilia occasionally lacking); calyptra cucullate.

Leaves secund, with margins recurved below, serrulate above;
  alar cells oblong and inflated in a single
  transverse row . . . . 1. *Brotherella*
Leaves straight; margins erect and entire; alar cells
  subquadrate in numerous rows . . . . . 2. *Heterophyllium*

1. *Brotherella* Loeske

Plants slender to medium-sized, in green or yellowish, ± glossy,

[174]

flat mats; stems creeping, freely branched; leaves ± complanate and often secund, concave, oblong-ovate or oblong-lanceolate, slenderly acuminate; margins recurved below, serrulate above; costa none; cells linear above, yellow at the insertion, larger, oblong, and inflated at the basal angles in a single transverse row; setae elongate; capsules erect or ± inclined; peristome double, the cilia ± rudimentary.

1. Plants very slender, scarcely complanate-foliate;
   capsules erect and symmetric . . . . . 3. *B. tenuirostris*
1. Plants somewhat larger, complanate-foliate;
   capsules ± inclined and asymmetric 2
   2. Plants light-green or yellowish, very glossy;
      operculum half as long as the urn . . . . 1. *B. recurvans*
   2. Plants darker, less shiny; operculum
      about as long as the urn . . . . 2. *B. delicatula*

1. *Brotherella recurvans* (Mx.) Fleisch. — Plants of moderate size, in very glossy, green or yellow-green, flat mats; leaves complanate and falcate-secund, ovate-lanceolate, slenderly acuminate, serrate above; 4–8 cells at basal margins abruptly inflated, above them at the margins a few, small, subquadrate cells; setae 1–2 cm. long; capsules ± inclined, 1.5–2 mm. long; operculum about 1/2 as long as the urn; spores in late autumn. — On rotten logs, humus, bases of trees, etc. in moist woods. Widespread in eastern North America.

ILLUSTRATIONS: Figure 131. Conard, *How to Know the Mosses* (ed. 2), fig. 264. Grout, *Mosses with Hand-Lens and Microscope*, fig. 192 (as *Hypnum*). Jennings, *Mosses of Western Pennsylvania* (ed. 2), Pl. 51 (as *Stereodon*).

Upper Peninsula: Chippewa Co., *Steere.* Gogebic Co., *Bessey.* Marquette Co., *Nichols.* Ontonagon Co., *Darlington & Bessey.*
Lower Peninsula: Alpena Co., *Robinson & Wells.* Cheboygan Co., *Nichols.* Emmet Co., *Wheeler.* Leelanau Co., *Darlington.* Oscoda Co., *Parmelee.* Washtenaw Co., *Steere.*

2. *Brotherella delicatula* (James) Fleisch. — Plants smaller, darker, less glossy, with less constantly falcate-secund leaves; setae and capsules shorter; operculum as long as the urn. — On rotten wood, soil, and bases of trees in moist woods. Widespread in eastern North America, especially at higher elevations.

ILLUSTRATION: Grout, *Mosses with Hand-Lens and Microscope*, Pl. 81 (as *Hypnum laxepatulum*).

Upper Peninsula: Marquette Co., *Nichols.* Ontonagon Co., *Nichols & Steere.*

3. *Brotherella tenuirostris* (Bruch & Schimp.) Broth. — Plants dark- or yellow-green, not or somewhat shiny; leaves ± secund but

scarcely complanate, oblong-lanceolate, slenderly acuminate, serrate above; alar cells moderately inflated; setae about 1 cm. long; capsules erect and symmetric; operculum about 1/2 as long as the urn: spores in early spring. — On rotten wood, moist rocks, and bases of trees in woods. Widespread in eastern United States.

ILLUSTRATIONS: Jennings, *Mosses of Western Pennsylvania* (ed. 2), Pl. 51 (as *Stereodon*). Welch, *Mosses of Indiana*, fig. 198.

Lower Peninsula: Eaton Co. (Grand Ledge), *Darlington*.

2. *Heterophyllium* (Schimp.) Kindb.

Plants medium-sized, in green or yellowish, dull or shiny mats; stems creeping, pinnately branched; paraphyllia present, few; leaves erect or erect-spreading, sometimes secund, ± concave, lanceolate or ovate-lanceolate, acuminate; costa very short or lacking; cells smooth, linear, yellowish at the insertion, subquadrate and somewhat inflated in well-marked alar groups; setae elongate, smooth; capsules cylindric, suberect or inclined, ± curved, symmetric or nearly so; peristome double, perfect.

Heterophyllium haldanianum (Grev.) Kindb. — Leaves loosely imbricate, not secund, smooth, entire; capsules suberect or inclined, long-cylindric, somewhat curved; spores in late autumn or winter. — On dry logs and stumps, usually in partial shade and sometimes on rock, soil, humus, and bases of trees in dry or moist forests. Widespread in eastern North America; Europe and Asia.

ILLUSTRATIONS: Figure 132. Conard, *How to Know the Mosses* (ed. 2), fig. 275. Jennings, *Mosses of Western Pennsylvania* (ed. 2), Pl. 50 (as *Stereodon*).

Upper Peninsula: Alger Co., *Wheeler*. Chippewa Co., *Thorpe*. Gogebic Co., *Golley*. Houghton Co., *Parmelee*. Marquette Co., *Nichols*. Ontonagon Co., *Darlington & Bessey*.
Lower Peninsula: Alpena Co., *Robinson & Wells*. Charlevoix Co., *Parmelee*. Cheboygan Co., *Wheeler*. Clinton Co., *Parmelee*. Emmet Co., *Wheeler*. Ingham and Leelanau counties, *Darlington*. Missaukee Co., *Beattie*. Washtenaw Co., *Kauffman*.

HYPNACEAE

Plants small to robust, in loose or dense, sometimes shiny mats; stems creeping (or sometimes ascending), freely branched and often regularly pinnate; leaves occasionally somewhat complanate, mostly homomallous to secund or falcate-secund, mostly lanceolate to ovate, acuminate; costa short and double or lacking; cells elongate, smooth (or sometimes papillose at the back because of projecting angles), mostly differentiated at the basal angles; setae

[176]

elongate; capsules erect or more commonly inclined to horizontal, straight or curved and asymmetric; peristome double (or the endostome rarely lacking), usually perfect; calyptra cucullate.

1. Plants bearing dense clusters of brood-branches
   at the ends of ordinary branches . . . . . 5. *Platygyrium*
1. Plants not bearing brood-branches 2
   2. Leaves strongly plicate . . . . 6. *Ptilium*
   2. Leaves smooth 3
3. Stem leaves decurrent, papillose at back because
   of projecting cell angles . . . . 7. *Ctenidium*
3. Stem leaves not decurrent or papillose 4
   4. Leaves strongly homomallous to secund
      or falcate-secund 5
   4. Leaves straight (or sometimes slightly
      homomallous or secund) 6
5. Capsules erect and symmetric; branches curved when dry;
   leaves homomallous . . . . 4. *Pylaisia*
5. Capsules inclined and asymmetric; branches straight;
   leaves secund or falcate-secund . . . . . . 1. *Hypnum*
   6. Plants minute; leaves 0.6 mm. long, or less;
      median cells short, oblong-hexagonal; quadrate
      alar cells few or fairly numerous, not
      very conspicuous . . . . 3. *Amblystegiella*
   6. Plants rather small; leaves about 0.8 mm. long;
      median cells elongate, rhomboidal; quadrate
      alar cells numerous . . . . 2. *Homomallium*

1. *Hypnum* Hedw.

Plants slender to robust, in soft, green or yellow or brownish mats or tufts, often regularly pinnate; leaves crowded, secund or falcate-secund, sometimes somewhat complanate, ovate-lanceolate, acuminate, not or slightly decurrent; costa short and double or lacking; cells linear, smooth, generally thick-walled and porose at the insertion, small and subquadrate or larger and inflated at the basal angles; setae elongate, smooth; capsules suberect to horizontal, ± curved and asymmetric; peristome double, perfect.

1. Branching regularly pinnate, sometimes frondose 2
1. Branching irregularly pinnate 7
   2. Alar cells conspicuously inflated, hyaline
      and thin-walled . . . . 5. *H. lindbergii*
   2. Alar cells not or only slightly inflated,
      mostly small and subquadrate 3
3. Capsules long-cylindric, nearly erect and
   symmetric . . . . 1. *H. imponens*
3. Capsules ovoid-cylindric, strongly inclined, asymmetric 4
   4. Quadrate alar cells very numerous 5
   4. Quadrate alar cells few or none 6
5. Leaves subentire, with a few larger cells at the
   extreme basal angles; dioicous . . . . 2. *H. cupressiforme*

5. Leaves serrate, without enlarged cells at the
   basal angles; autoicous . . . . 7. *H. reptile*
6. Capsules plicate when dry . . . . . 3 *H. curvifolium*
6. Capsules not plicate . . . . 4. *H. fertile*
7. Alar cells large, hyaline, and inflated;
   cortical cells large and hyaline 8
7. Alar and cortical cells not as above 9
   8. Leaves strongly falcate secund, not
      particularly complanate . . . . . . 5. *H. lindbergii*
   8. Leaves ± secund at the tips,
      strongly complanate . . . . 6. *H. pratense*
9. Quadrate alar cells few (less than
   6 along the margins) . . . . 8. *H. pallescens*
9. Quadrate alar cells very numerous (6–15 along
   the margins) 10
   10. Leaves serrulate nearly to the base . . . . 7. *H. reptile*
   10. Leaves entire or serrulate at
       apex only . . . . 2. *H. cupressiforme*

1. *Hypnum imponens* Hedw. — Plants dark- or yellow-green above, brownish below, regularly pinnate; stem leaves falcate-secund, broadly ovate, gradually acuminate, somewhat rounded at insertion, serrulate; basal cells incrassate and orange-brown, a relatively few alar cells quadrate (4–6 along the margins) with larger and slightly inflated cells at the extreme angles (3–4 along the margins); branch leaves smaller and narrower; capsules cylindric, erect and symmetric or slightly curved; spores in autumn or early winter. — On rotten logs or sometimes on rocks, humus, or tree bases. Widespread in eastern North America; Europe, Azores, and Asia.

ILLUSTRATIONS: Figure 133. Conard, *How to Know the Mosses* (ed. 2), fig. 284. Grout, *Mosses with Hand-Lens and Microscope*, fig. 185. Jennings, *Mosses of Western Pennsylvania* (ed. 2), Pl. 49 (as *Stereodon*). Welch, *Mosses of Indiana*. fig. 191.

Upper Peninsula: Chippewa Co., *Steere*. Houghton Co., *Parmelee*. Keweenaw Co., *Cooper*. Marquette Co., *Nichols*.
Lower Peninsula: Alpena Co., *Robinson & Wells*. Cheboygan Co., *Nichols*. Charlevoix Co., *Parmelee*. Ingham and Leelanau counties, *Darlington*. Washtenaw Co., *Kauffman*.

2. *Hypnum cupressiforme* Hedw. — Plants usually irregularly branched; leaves ovate-lanceolate, acuminate, falcate-secund, entire or slightly serrulate above; alar cells dense and quadrate in fairly large groups (6–10 cells along the margins and only 2–3 slightly inflated cells at the extreme basal angles); capsules curved and inclined; spores in autumn to early winter. — On calcareous rock or soil, often in dry, exposed habitats. Widespread in Canada, Alaska, and northern United States and far southward at higher elevations; nearly cosmopolitan.

[178]

ILLUSTRATIONS: Conard, *How to Know the Mosses* (ed. 2), fig. 283a–e. Grout, *Mosses with Hand-Lens and Microscope*, fig. 186. Welch, *Mosses of Indiana*, fig. 192.

Upper Peninsula: Alger Co., *Nichols*. Mackinac Co., *Wheeler*.
Lower Peninsula: Cheboygan Co., *Nichols*. Huron Co., *Schnooberger*. Leelanau Co., *Darlington*. Washtenaw Co., *Kauffman*.

3. **Hypnum curvifolium** Hedw. — Plants rather large, green or yellow-green above, brown below, pinnately branched; stem leaves falcate-secund in 2 rows, triangular-ovate, gradually acuminate, entire or serrulate near the apex, rounded to the insertion, somewhat decurrent; cells of the decurrencies enlarged, thin-walled and hyaline, above them a few, short, oblong cells; capsules curved and inclined, plicate when dry and empty; spores in spring. — On rotten wood, rocks, or soil in woods. Widespread in eastern North America; Japan.

ILLUSTRATIONS: Conard, *How to Know the Mosses* (ed. 2), fig. 281. Grout, *Mosses with Hand-Lens and Microscope*, fig. 187. Jennings, *Mosses of Western Pennsylvania* (ed. 2), Pl. 50 (as *Stereodon*). Welch, *Mosses of Indiana*, fig. 193.

Upper Peninsula: Keweenaw Co., *Cooper*.
Lower Peninsula: Cheboygan and Ingham counties, *Wheeler*. Leelanau Co., *Darlington*. Washtenaw Co., *Kauffman*.

4. **Hypnum fertile** Sendt. — Plants light- or yellow-green, brownish below, regularly pinnate; stem leaves falcate-secund, oblong-lanceolate, gradually and slenderly acuminate, serrulate near the apex, rounded at base, not decurrent; a few cells at the extreme basal angles slightly oblong and inflated, above them 3–4 denser, subquadrate cells; capsules curved and inclined; spores in early summer. — On rotten logs, sometimes also on soil, rocks, or bases of trees in woods. Across Canada, south to Tennessee; Europe and Asia.

ILLUSTRATIONS: Grout, *Mosses with Hand-Lens and Microscope*, fig. 189. Jennings, *Mosses of Western Pennsylvania* (ed. 2), Pl. 49 (as *Stereodon*).

Upper Peninsula: Keweenaw Co., *Steere*. Ontonagon Co., *Nichols & Steere*.
Lower Peninsula: Ingham and Oakland counties, *Darlington*.

5. **Hypnum lindbergii** Mitt. — Plants light- or yellow-green, shiny, irregularly to subpinnately branched; cortical cells large and hyaline; leaves falcate-secund in 2 rows, oblong-ovate, broadly acuminate, decurrent, entire or serrulate at apex; alar cells abruptly inflated, bordered by smaller, subquadrate cells; capsules curved and inclined, plicate when dry and empty; spores in June. — On wet soil or other substrata in swamps and meadows. Across the continent, south to Washington, Colorado, and Florida; Europe and Asia.

[179]

ILLUSTRATIONS: Conard, *How to Know the Mosses* (ed. 2), fig. 276 (as *H. patientiae*). Grout, *Mosses with Hand-Lens and Microscope*, fig. 188 (as *H. patientiae*). Jennings, *Mosses of Western Pennsylvania* (ed. 2), Pl. 50 (as *Stereodon patientiae*). Welch, *Mosses of Indiana*, fig. 194.

Upper Peninsula: Chippewa and Keweenaw counties, *Steere*. Ontonagon Co., *Nichols & Steere*.
Lower Peninsula: Alpena Co., *Hall*. Cheboygan Co., *Nichols*. Clinton Co., *Parmelee*. Ingham and Leelanau counties, *Darlington*. Oscoda Co., *Parmelee*. Washtenaw Co., *Steere*.

6. *Hypnum pratense* Koch — Closely related to *H. lindbergii* but irregularly branched; leaves complanate, slightly secund at the tips; alar cells less sharply differentiated, not bordered by smaller, sub-quadrate cells; spores in spring. — In swamps and wet meadows. Across the continent, south to Pennsylvania, Colorado, and British Columbia; Europe and Japan.

ILLUSTRATION: Jennings, *Mosses of Western Pennsylvania* (ed. 2), Pl. 70 (as *Stereodon*).

Upper Peninsula: Chippewa Co., *Steere*. Marquette Co., *Nichols*. Keweenaw Co., *Allen & Stuntz*.
Lower Peninsula: Cheboygan Co., *Nichols*. Oakland Co., *Darlington*. Washtenaw Co., *Steere*.

7. *Hypnum reptile* Mx. — Small, dark-green, ± regularly pinnate plants; leaves falcate-secund, ovate, slenderly acuminate, slightly decurrent, serrate above, entire or serrulate below; alar cells dense and subquadrate in rather large groups (10–20 cells along the margins); capsules curved and inclined; spores in summer. — On bases of trees, rotten logs, and rocks in woods. Nova Scotia to Alaska, south at higher elevations to Arizona and North Carolina; Europe and Asia.

ILLUSTRATIONS: Conard, *How to Know the Mosses* (ed. 2), fig. 285. Grout, *Mosses with Hand-Lens and Microscope*, fig. 190. Jennings, *Mosses of Western Pennsylvania* (ed. 2), Pl. 49 (as *Stereodon*). Welch, *Mosses of Indiana*, fig. 195.

Upper Peninsula: Chippewa Co., *Steere*. Gogebic Co., *Golley*. Houghton Co., *Hyypio*. Keweenaw Co., *Allen & Stuntz*. Marquette Co., *Nichols*. Ontonagon Co., *Darlington*.
Lower Peninsula: Alpena Co., *Robinson & Wells*. Cheboygan and Emmet counties, *Nichols*. Leelanau Co., *Darlington*. Oscoda Co., *Parmelee*. Washtenaw Co., *Steere*.

8. *Hypnum pallescens* (Hedw.) BSG — Very similar to *H. reptile;* plants less regularly pinnate; leaves more slenderly acuminate, less falcate, secund, less strongly serrate, with fewer quadrate alar cells; capsules shorter, nearly erect and symmetric. — On bark of conifers, rotten wood, etc. Across the continent and southward in the mountains to Tennessee; Europe.

[180]

ILLUSTRATION: Grout, *Moss Flora of North America*, vol. 3, Pl. 33C.

Upper Peninsula: Alger Co., *Wheeler*. Keweenaw Co., *Allen & Stuntz*. Mackinac Co., *Kauffman*. Marquette Co., *Nichols*. Ontonagon Co., *Nichols & Steere*.
Lower Peninsula: Cheboygan Co., *Wheeler*. Leelanau Co., *Darlington*.

2. *Homomallium* (Schimp.) Loeske

Rather slender or medium-sized plants in green or yellowish, rather shiny mats; stems creeping, freely branched; branches short, erect or curved; leaves erect-spreading or homomallous, oblong-lanceolate, acuminate, entire or ± serrulate at apex; costa none, short and double, or weak and single; median cells elongate; alar cells quadrate in numerous rows; capsules inclined, curved, oblong-cylindric; peristome double, perfect.

Homomallium adnatum (Hedw.) Broth. — Leaves erect-spreading when moist, loosely erect when dry, not at all secund, oblong-ovate, abruptly short-acuminate, entire, about 0.8 mm. long; costa short and double (or rarely single), sometimes lacking; quadrate alar cells sometimes extending 1/3 up the leaf; spores in summer. — On rocks or bases of trees and stumps in woods. Widespread in eastern North America; Asia.

ILLUSTRATIONS: Figure 134. Jennings, *Mosses of Western Pennsylvania* (ed. 2), Pl. 43. Welch, *Mosses of Indiana*, fig. 201.

Upper Peninsula: Ontonagon Co. (Porcupine Mts.), *Darlington*.
Lower Peninsula: Cheboygan Co., *Nichols*. Clinton Co., *Parmelee*. Ingham Co., *Darlington*. Washtenaw Co., *Steere*.

3. *Amblystegiella* Loeske

Very small plants in dull or somewhat shiny mats, sometimes mixed with other mosses; stems creeping, irregularly branched; leaves very small, erect-spreading or sometimes weakly secund, lanceolate or lance-subulate, entire; costa none or very short and weak; cells smooth, short, oblong-hexagonal or shortly rhomboidal, ± subquadrate and poorly differentiated at the basal angles; setae elongate; capsules mostly erect and symmetric or nearly so; peristome double and perfect (or the cilia lacking or rudimentary).

1. Plants growing on bark at base of trees; leaves up
   to about 0.6 mm. long . . . . 1. *A. subtilis*
1. Plants on rock; leaves less than 0.4 or 0.5 mm. long 2
   2. Monoicous; capsules curved and inclined;
      quadrate alar cells fairly numerous . . . 2. *A. confervoides*
   2. Dioicous; capsules suberect; quadrate
      alar cells few or none . . . . 3. *A. sprucei*

[181]

1. *Amblystegiella subtilis* (Hedw.) Loeske — Small, dark-green plants; leaves appressed when dry, narrowly lanceolate, slenderly acuminate, up to 0.6 mm. long, entire; upper cells oblong-hexagonal (2–3:1); quadrate alar cells numerous; capsules suberect; spores in August and September. — On tree bases. Eastern North America, south to Pennsylvania and Illinois; Europe and Asia.

ILLUSTRATIONS: Conard, *How to Know the Mosses* (ed. 2), fig. 260a–d. *Mosses with Hand-Lens and Microscope*, fig. 201, 202 (3a–d).

Upper Peninsula: Baraga Co., *Nichols & Steere*. Gogebic Co., *Golley*. Houghton Co., *Parmelee*. Keweenaw Co., *Povah*. Ontonagon Co., *Nichols & Steere*.
Lower Peninsula: Alpena Co., *Robinson & Wells*. Cheboygan Co., *Nichols*. Gratiot and Huron counties, *Schnooberger*. Leelanau Co., *Darlington*. Newaygo and Otsego counties, *Schnooberger*.

2. *Amblystegiella confervoides* (Brid.) Loeske — Small, dark-green plants; leaves erect or erect-spreading, ovate or ovate-lanceolate, acute or stoutly acuminate, subentire, ecostate; cells about 3:1; quadrate alar cells rather numerous; autoicous; capsules ± asymmetric, inclined; spores in late summer or autumn. — On rocks, especially limestone. Southeastern Canada and northeastern United States; Rocky Mts.; Europe.

ILLUSTRATIONS: Figure 135. Conard, *How to Know the Mosses* (ed. 2), fig. 260e, f. Grout, *Mosses with Hand-Lens and Microscope*, fig. 202a–d. Welch, *Mosses of Indiana*, fig. 202.

Upper Peninsula: Mackinac Co., *Nichols*.
Lower Peninsula: Washtenaw Co., *Steere*.

3. *Amblystegiella sprucei* (Bruch) Loeske — Very small, bright-green plants, often growing in mixture with other mosses; leaves spreading, often subsecund, ovate-lanceolate, slenderly acuminate, serrulate; upper cells 3–6:1; alar cells scarcely differentiated; dioicous; capsules erect and symmetric or nearly so; spores in summer. — On soil, rock, and other substrata in sheltered niches, especially in crevices of cliffs. Across Canada and Alaska, south to New England and New Mexico; Europe and Asia.

ILLUSTRATION: Grout, *Mosses with Hand-Lens and Microscope*, fig. 202 (2a–d).

Lower Peninsula: Alpena Co., *Robinson & Wells*. Cheboygan Co., *Fulford*.

4. *Pylaisia* BSG

Plants rather small or medium-sized in flat, sometimes shiny, green or yellowish mats; stems creeping, freely branched; branches short, ascending, often curved; leaves crowded, loosely appressed when dry, usually homomallous, concave, ovate- or oblong-lanceolate, short- or long-acuminate, entire or nearly so; costa none or very short and double; cells smooth, linear or rhomboidal above, quadrate in several rows at the basal angles; autoicous; setae elongate; capsules ovoid- or oblong-cylindric, erect and symmetric; peristome double, the endostome often adhering wholly or in part to the exostome, basal membrane high, cilia of endostome mostly rudimentary.

1. Segments of endostome free; spores 10–12 μ 2
1. Segments partly or wholly adherent to the
    exostome; spores 18 μ or more 3
    2. Quadrate alar cells numerous (about
       10–15 along the margins) . . . . . 2. *P. subdenticulata*
    2. Quadrate alar cells fewer (3–9 at
       the margins) . . . . 1. *P. polyantha*
3. Segments partly adherent to the teeth;
    spores 18–26 μ . . . . 3. *P. selwynii*
3. Segments wholly adherent to the teeth;
    spores 24–30 μ . . . . 4. *P. intricata*

1. *Pylaisia polyantha* (Hedw.) BSG — Yellow-green, glossy plants; branch leaves crowded, ± secund, ovate-lanceolate, long-acuminate; quadrate alar cells relatively few (3–9 along the margins); segments free; spores 10–12 μ, maturing in autumn and winter. — On bark of treees. Across the continent and south to the northern United States and Arizona; Europe, North Africa, and Asia.

ILLUSTRATIONS: Figure 136. Conard, *How to Know the Mosses* (ed. 2), fig. 288a–e. Grout, *Moss Flora of North America*, vol. 3, Pl. 35G.

Upper Peninsula: Chippewa Co., *Steere.* Keweenaw Co., *Allen & Stuntz.*
Lower Peninsula: Cheboygan Co., *Nichols.* Van Buren Co., *Kauffman.*

2. *Pylaisia subdenticulata* BSG — Branch leaves erect-spreading, ovate-lanceolate, gradually long-acuminate, subdenticulate at apex; quadrate alar cells numerous (10–15 at the margins); segments free; spores 10–12 μ, maturing in summer or early autumn. — On bark of trees. Across the continent, south to North Carolina and New Mexico.

ILLUSTRATIONS: Conard, *How to Know the Mosses* (ed. 2), fig. 288f. Grout, *Moss Flora of North America*, vol. 3, Pl. 37B.

Upper Peninsula: Chippewa Co., *Steere.* Keweenaw Co., *Allen & Stuntz.* Ontonagon Co., *Nichols & Steere.*
Lower Peninsula: Cheboygan and Emmet counties, *Nichols.* Leelanau Co., *Darlington.*

3. *Pylaisia selwynii* Kindb. — Branch leaves homomallous, ovate-lanceolate, ± long-acuminate, entire or subdenticulate at apex; quadrate alar cells numerous (15–20 at the margins); segments adherent to the teeth for 2/3 their length; spores 18–26 μ, maturing in autumn. — On bark of trees. Widespread in eastern North America; Arizona; Europe and Asia.

ILLUSTRATIONS: Conard, *How to Know the Mosses* (ed. 2), fig. 287. Grout, *Mosses with Hand-Lens and Microscope*, Pl. 85 (as *P. schimperi.*) Jennings, *Mosses of Western Pennsylvania* (ed. 2), Pl. 37. Welch, *Mosses of Indiana*, fig. 204.

Upper Peninsula: Chippewa Co., *Steere*. Keweenaw Co., *Hermann*. Ontonagon Co., *Nichols & Steere*.
Lower Peninsula: Alpena Co., *Robinson & Wells*. Cheboygan Co., *Nichols*. Clinton Co., *Parmelee*. Leelanau Co., *Darlington*. Macomb Co., *Cooley*. Oscoda Co., *Parmelee*.

4. *Pylaisia intricata* (Hedw.) BSG — Branch leaves homomallous, ovate-lanceolate, ± long-acuminate, entire or denticulate; quadrate alar cells few (less than 10 at the margins); endostome segments wholly adherent to the exostome; spores 24–30 μ, maturing in late summer or early autumn. — On bark of trees. Widespread in eastern North America; Arizona.

ILLUSTRATIONS: Grout, *Mosses with Hand-Lens and Microscope*, fig. 209. Jennings, *Mosses of Western Pennsylvania* (ed. 2), Pl. 37. Welch, *Mosses of Indiana*, fig. 205.

Upper Peninsula: Alger Co., *Wheeler*. Mackinac Co., *Nichols*.
Lower Peninsula: Van Buren Co., *Kauffman*.

5. *Platygyrium* BSG

Rather small, dark-green, yellowish, or brownish, glossy plants in flat mats, irregularly branched; leaves crowded, appressed when dry, spreading when moist, ovate or oblong-lanceolate, short- or long-acuminate, entire; costa none; cells smooth, rhomboidal at the apex, linear below, quadrate in several rows at the basal angles; dioicous; setae elongate; capsules erect and symmetric or slightly curved; peristome double, basal membrane low, cilia none.

Platygyrium repens (Brid.) BSG — Plants dark- or blackish-green with a golden sheen; many branches ending in dense clusters of tiny brood-branches; spores in autumn. — On bogs, stumps, bark of trees, and (rarely) rocks. Widespread in eastern North America; Europe.

ILLUSTRATIONS: Figure 137. Conard, *How to Know the Mosses* (ed. 2), fig. 261. Grout, *Mosses with Hand-Lens and Microscope*, fig. 208. Jennings, *Mosses of Western Pennsylvania* (ed. 2), Pl. 37. Welch, *Mosses of Indiana*, fig. 206.

Upper Peninsula: Chippewa Co., *Steere.* Gogebic Co., *Golley.* Keweenaw Co., *Parmelee.* Marquette Co., *Nichols.* Ontonagon Co., *Darlington.*

Lower Peninsula: Alpena Co., *Hall.* Cass Co., *Darlington.* Cheboygan and Emmet counties, *Nichols.* Genesee, Ingham, Leelanau, and Monroe counties, *Darlington.* St. Clair Co., *Gilly & Parmelee.* St. Joseph Co., *Darlington.*

6. *Ptilium* (Sull.) DeNot.

Robust plants in loose, yellow-green or golden mats; stems ascending, at least above, sometimes forked, regularly pinnate, forming oblong-triangular fronds, hooked at the tips; branches also hooked; leaves crowded, strongly falcate-secund, deeply plicate, ovate, slenderly long-acuminate, serrulate above; costa very short and double or lacking; cells smooth, linear, not much differentiated at the basal angles; dioicous; setae long, smooth; capsules inclined, curved-cylindric; operculum hemispheric and apiculate; peristome double, perfect. — A genus of one species, *P. crista-castrensis* (Hedw.) DeNot. —On humus and old logs in moist woods. Across the continent, south to North Carolina and Iowa; Europe and Asia.

ILLUSTRATIONS: Figure 138. Conard, *How to Know the Mosses* (ed. 2), fig. 279 (as *Hypnum*). Grout, *Mosses with Hand-Lens and Microscope*, fig. 181, 182 (*as Hypnum*). Jennings, *Mosses of Western Pennsylvania* (ed. 2), Pl. 49.

Upper Peninsula: Alger Co., *Wheeler.* Chippewa Co., *Thorpe.* Delta Co., *Wheeler.* Houghton and Keweenaw counties, *Parmelee.* Mackinac Co., *Wheeler.* Marquette Co., *Nichols.* Ontonagon Co., *Darlington.*

Lower Peninsula: Alpena Co., *Hall.* Cheboygan Co., *Nichols.* Emmet Co., *Wheeler.* Leelanau Co., *Darlington.* Washtenaw Co., *Kauffman.*

7. *Ctenidium* (Schimp.) Mitt.

Plants rather small, in soft, green or yellowish, rather glossy mats; stems generally prostrate, ± pinnate; stem and branch leaves somewhat differentiated; stem leaves wide-spreading or secund, smooth or somewhat plicate, mostly abruptly lance-subulate from an ovate-cordate, broadly decurrent base, toothed all around; costa short and double or none; cells linear, usually papillose at back because of projecting ends, subquadrate in small alar groups; branch leaves smaller and ovate-lanceolate; setae elongate, smooth or somewhat roughened; capsules inclined, oblong-ovoid; peristome double, perfect.

Ctenidium molluscum (Hedw.) Mitt. — Plants glossy, golden-green, usually regularly pinnate and frondose; leaves imbricate when dry, falcate-secund, plicate; cells somewhat papillose; dioicous. — On moist, shaded soil, humus, and rocks. Widespread in eastern North America; Europe, Asia, and North Africa.

[185]

ILLUSTRATIONS: Figure 139. Conard, *How to Know the Mosses* (ed. 2), fig. 280. Grout, *Mosses with Hand-Lens and Microscope*, fig. 183 (as *Hypnum*). Jennings, *Mosses of Western Pennsylvania* (ed. 2), Pl. 46. Welch, *Mosses of Indiana*, fig. 196 (as *Hypnum*).

Lower Peninsula: Clare, Gratiot, and Isabella counties, *Schnooberger*.

HYLOCOMIACEAE

Robust, often rigid and shiny plants in loose mats or tufts; stems prostrate or erect-ascending, sparsely or pinnately to ± tripinnately branched, sometimes frondose; paraphyllia none, few, or many; leaves broadly ovate to oblong-lanceolate, acute to gradually or abruptly long-acuminate, sometimes plicate or rugose; costa single or double, often weak, sometimes lacking; cells prosenchymatous, sometimes papillose at back because of projecting cell angles, not or somewhat differentiated at basal angles; dioicous; setae elongate, smooth; capsules usually inclined, generally asymmetric; peristome double, perfect; calyptra cucullate.

1. Paraphyllia abundant . . . . 1. *Hylocomium*
1. Paraphyllia lacking or very few 2
  2. Leaves secund and rugose; costa single, reaching
    the leaf middle or beyond . . . . . . 3. *Rhytidium*
  2. Leaves spreading or squarrose, often plicate
    but not rugose; costa double, often strong . 2. *Rhytidiadelphus*

1. *Hylocomium* BSG

Robust plants in loose, yellowish or olive-green mats or tufts; stems arcuate and procumbent, irregularly pinnate to regularly 2–3-pinnate and often frondose from a stipitate base; paraphyllia very abundant; stem leaves lightly plicate, oblong-ovate, short- or long-acuminate, serrulate all around; costa single or short and double; cells linear, smooth or minutely papillose at back because of projecting ends, not differentiated at basal angles; setae long; capsules oblong-ovoid, horizontal; operculum rostrate.

1. Costa single, ending near the leaf middle
  (rarely forked) . . . . 3. *H. pyrenaicum*
1. Costa double 2
  2. Branching irregular or ± pinnate;
    costae elongate . . . . 2. *H. umbratum*
  2. Branching very regularly 2–3 pinnate;
    costae short . . . . 1. *H. splendens*

1. *Hylocomium splendens* (Hedw.) BSG — Plants regularly 2–3-pinnate and frondose from a stipitate base, becoming layered because of annual growth from stoloniform innovations; stem leaves broadly ovate, abruptly acuminate, minutely papillose at back be-

[186]

cause of projecting cell ends; costa double, short, occasionally nearly 1/2 the leaf length; branch leaves small, ovate, acute; spores in spring. — On humus, soil, rock, etc. in moist, shady places. Across the continent, south to North Carolina, Colorado, and California; Europe, Asia, Africa, and New Zealand.

ILLUSTRATIONS: Figure 140. Conard, *How to Know the Mosses* (ed. 2), fig. 172. Grout, *Mosses with Hand-Lens and Microscope*, fig. 140 (as *H. proliferum*). Jennings, *Mosses of Western Pennsylvania* (ed. 2), Pl. 47.

Upper Peninsula: Alger Co., *Wheeler*. Chippewa Co., *Steere*. Keweenaw Co., *Parmelee*. Mackinac Co., *Kauffman*. Marquette Co., *Nichols*.

Lower Peninsula: Alpena, Cheboygan, and Emmet counties, *Wheeler*, Leelanau Co., *Darlington*. Washtenaw Co., *Steere*.

2. Hylocomium umbratum (Hedw.) BSG

2. *Hylocomium umbratum* (Hedw.) BSG — Resembling *H. splendens* but the annual layers less apparent, rather irregularly 1–2-pinnate but not in 1 plane, giving a bushy appearance; stem leaves triangular-cordate, acute to long-acuminate, long-decurrent, serrate, deeply plicate; costa double, ending near the leaf middle or beyond; cells strongly papillose at back; spores in winter or spring. — On stones, old logs, and humus in moist, shady places. Canada and Alaska, south to North Carolina; Europe and Asia.

ILLUSTRATIONS: Conard, *How to Know the Mosses* (ed. 2), fig. 198a–d. Grout, *Mosses with Hand-Lens and Microscope*, Pl. 61. Jennings, *Mosses of Western Pennsylvania* (ed. 2), Pl. 48.

Upper Peninsula: Marquette Co. (Huron Mts.) *Nichols*.
Lower Peninsula: Cheboygan Co., *Steere*.

3. Hylocomium pyrenaicum (Spruce) Lindb.

3. *Hylocomium pyrenaicum* (Spruce) Lindb. — Robust, dark- or yellow-green plants; secondary stems ascending, irregularly branched; stems and branches turgid at apex; leaves broadly ovate, acute or broadly acuminate, serrate above, plicate, not or slightly decurrent; costa single, about 1/2 the leaf length; cells smooth; spores in winter or early spring. — On logs and rocks in moist woods. Across the continent, south to New England and Minnesota; Europe and Asia.

ILLUSTRATION: Grout, *Moss Flora of North America*, vol. 3, Pl. 31B.

Upper Peninsula: Chippewa Co., *Steere*. Gogebic Co., *Bessey*. Keweenaw Co., *Steere*. Marquette Co., *Nichols*. Ontonagon Co., *Darlington*.

2. Rhytidiadelphus (Lindb.) Warnst.

Coarse, robust plants in loose, green or yellowish tufts; stems prostrate to erect, sparsely to pinnately branched; paraphyllia none or very few; leaves erect and ovate or cordate at base and gradually or abruptly narrowed to a squarrose or falcate acumen, mostly striate or plicate, serrulate to serrate; costa double, short or sometimes reaching the leaf middle; cells linear, smooth or spinose at back because of projecting ends; setae long; capsules inclined, oblong-ovoid, asymmetric; peristome double, perfect.

Stem leaves spinose at back, short-acuminate . . . 1. *R. triquetrus*
Stem leaves smooth at back, slenderly
    acuminate . . . . 2. *R. squarrosus*

1. *Rhytidiadelphus triquetrus* (Hedw.) Warnst. — Very robust, green or yellow-green plants; stems erect or ascending, irregularly or subpinnately branched; leaves wide-spreading, stiff and scarious, plicate, cordate-triangular, broadly acuminate, rounded-auriculate at base, dentate above, denticulate below; costa double, often extending 3/4 the leaf length; upper cells spinose-papillose because of projecting ends; spores in winter or spring. — On humus and rotten wood in dry or moist forests. Across North America, south to North Carolina, Missouri, and California; Europe and Asia.

ILLUSTRATIONS: Conard, *How to Know the Mosses* (ed. 2), fig. 168. Grout, *Mosses with Hand-Lens and Microscope*, fig. 141 (as *Hylocomium*). Jennings, *Mosses of Western Pennsylvania* (ed. 2), Pl. 47.

Upper Peninsula: Alger Co., *Wheeler*. Chippewa Co., *Steere*. Houghton and Keweenaw counties, *Parmelee*. Mackinac Co., *Wheeler*. Marquette Co., *Nichols*. Ontonagon Co., *Nichols & Steere*.

Lower Peninsula: Alpena Co., *Hall*. Cheboygan, Dickinson, and Emmet counties, *Wheeler*. Leelanau Co., *Darlington*. Washtenaw Co., *Kauffman*.

2. *Rhytidiadelphus squarrosus* (Hedw.) Warnst. — Plants green or yellow-green; stems erect or ascending from a prostrate base, freely branched; leaves sheathing at the cordate-ovate base, abruptly narrowed to a long, slender, squarrose-recurved acumen, not plicate, denticulate, especially above; costa double, rather short (scarcely 1/2 the leaf length); cells smooth; alar cells clearly differentiated; spores in winter or spring. — On soil, humus, logs, or rocks in wet, shady places. Across the continent, south at higher elevations to Tennessee; Europe, Macaronesia, and Asia.

ILLUSTRATIONS: Figure 141. Conard, *How to Know the Mosses* (ed. 2), fig. 289. Grout, *Mosses with Hand-Lens and Microscope*, fig. 142 (as *Hylocomium*).

Upper Peninsula: Chippewa Co., *Steere*. Keweenaw Co., *Hermann*.
Lower Peninsula: Cheboygan Co., *Nichols*.

3. *Rhytidium* (Sull.) Kindb.

Rather robust, turgid, yellowish to brownish plants in loose tufts, rigid and shiny when dry; stems prostrate to erect, irregularly to pinnately branched; paraphyllia none; leaves appressed with secund tips, both plicate and rugose, oblong-ovate, slenderly acuminate; margins narrowly recurved, toothed near the apex; costa single, ending near the leaf middle; cells linear, some projecting at the ends as sharp papillae; alar cells well differentiated, subquadrate; setae long; capsules inclined, oblong-cylindric, asymmetric; peristome double, perfect. — A genus of one species, *R. rugosum* (Hedw.) Kindb. — On bare, exposed rocks or thin soil or humus in dry, sometimes wooded slopes and bluffs. Greenland; arctic America to North Carolina and Mexico; Europe, North Africa, and Asia.

ILLUSTRATIONS: Figure 142. Conard, *How to Know the Mosses* (ed. 2), fig. 165. Grout, *Mosses with Hand-Lens and Microscope*, fig. 143. Jennings, *Mosses of Western Pennsylvania* (ed. 2), Pl. 66. Welch, *Mosses of Indiana*, fig. 190.

Upper Peninsula: Keweenaw Co. (Isle Royale), *Steere.*

BUXBAUMIACEAE

Very small, scattered to caespitose plants with exceedingly short stems and a ± persistent protonema; leaves very small and ephemeral or larger and persistent; capsules large, nearly sessile or on a moderately long, stout, rough seta, oblique and asymmetric, ovoid, ± ventricose; peristome single or double, the exostome (if present) of 1–several concentric rows of slender teeth, the endostome a pleated cone with a small opening at top; calyptra very small.

Leaves very small and ephemeral; plants scattered,
  appearing to consist only of sporophyte; setae present .   1. *Buxaumia*
Leaves persistent; plants in dense tufts; setae
  extremely short .   .   .   .   2. *Diphyscium*

1. *Buxbaumia* Hedw.

Small, scattered, inconspicuous mosses growing from a persistent protonema and consisting almost entirely of sporophyte; stems exceedingly short; leaves very small, few, disappearing early; setae elongate, stout, rough; capsules strongly inclined and very asymmetric, ovoid or elongate-ovoid, ventricose at base; peristome double, the exostome (sometimes rudimentary) consisting of 1–4 rows of short, irregular, papillose teeth, the endostome a pale, pleated cone open at the apex.

Capsules ovate in outline, flattened
  on the upper side .   .   .   .   1. *B. aphylla*
Capsules narrow and elongate, lanceolate in
  outline, not flattened above .   .   .   .   .   2. *B. subcylindrica*

[189]

1. *Buxbaumia aphylla* Hedw. — Capsules ovate in outline, flattened above, convex below and ventricose at base; spores in late autumn to spring. This moss, aptly called a "bug-on-a-stick," is often overlooked because of its resemblance to a small, stalked fungus. — A pioneer on bare sand or clay banks, sometimes on logs or stumps, in shaded or open places. Greenland; Yukon to Washington and southeastern Canada and northeastern United States to West Virginia; Europe, Asia, and New Zealand.

ILLUSTRATIONS: Figure 143. Conard, *How to Know the Mosses* (ed. 2), fig. 30a. Grout, *Mosses with Hand-Lens and Microscope*, fig. 19, Pl. 8. Jennings, *Mosses of Western Pennsylvania* (ed. 2), Pl. 30. Welch, *Mosses of Indiana*, fig. 63.

Upper Peninsula: Alger Co., *Thompson*. Mackinac Co., *Steere*.
Lower Peninsula: Cheboygan Co., *Nichols*. Emmet Co., *Steere*. Kent Co., *Kauffman*. Grand Traverse Co., *Steere*. Muskegon Co., *Darlington*. Oscoda Co., *Parmelee*. Presque Isle Co., *Steere*.

2. *Buxbaumia subcylindrica* Grout — Capsules elongate, lancecylindric, not flattened above, only slightly ventricose; spores in October and later. — On rotten wood in moist places. New England and New York to British Columbia.

ILLUSTRATION: Grout, *Moss Flora of North America*, vol. 1, Pl. 73.

Lower Peninsula: Kalamazoo Co. (Climax), *Becker*.

2. *Diphyscium* Mohr

Small, dark-green or blackish plants in dense, wide mats; leaves crisped when dry, lingulate, obtuse, often cucullate; costa vanishing below the apex; cells rounded-quadrate, incrassate, in 2 or more layers, pale and rectangular below; setae very short; capsules nearly sessile, exceeded by the awns of the perichaetial leaves, ovoid, ventricose, inclined and asymmetric; exostome none; endostome a whitish cone. — A small genus with one species in our area, *D. foliosum* (Hedw.) Mohr, which resembles a "scared rabbit in the grass." Its spores mature in early summer. — On moist, shaded banks in woods. Wide-ranging in eastern North America; Europe, North Africa, Madeira, Caucasus, Mexico, Guatemala, and the West Indies.

ILLUSTRATIONS: Figure 144. Conard, *How to Know the Mosses* (ed. 2), fig. 30b–d. Grout, *Mosses with Hand-Lens and Microscope*, fig. 20 (as *Webera sessilis*). Jennings, *Mosses of Western Pennsylvania* (ed. 2), Pl. 29. Welch, *Mosses of Indiana*, fig. 64.

Upper Peninsula: Chippewa Co., *Thorpe*. Houghton Co., *Parmelee*. Keweenaw Co., *Steere*. Mackinac Co., *Kauffman*. Ontonagon Co., *Nichols & Steere*.
Lower Peninsula: Cheboygan Co., *Nichols*. Eaton Co., *Parmelee*.

Usually robust, coarse, rigid plants with erect, simple or forked stems; leaves oblong-lanceolate from a broader, usually sheathing base, bearing many green lamellae on the upper surface of the blade (rarely also on the back), on the costa or the lamina, or both; costa single, well developed; cells subquadrate or short-oblong, with firm walls, sometimes differentiated at the margins; setae elongate; capsules erect or inclined, cylindric or angled; peristome consisting of 32–64 (rarely 16) solid, unjointed teeth attached to the expanded tip of the columella; calyptra large, cucullate, often hairy.

1. Leaves bordered; lamellae limited to
   the costa; calyptra smooth . . . . 1. *Atrichum*
1. Leaves not bordered; lamellae more numerous,
   on both costa and lamina; calyptra hairy 2
   2. Capsules cylindric . . . . 2. *Pogonatum*
   2. Capsules angled . . . . 3. *Polytrichum*

1. *Atrichum* P.-Beauv.

Plants small or medium-sized, dark-green or brownish, gregarious or loosely tufted; leaves often crisped and contorted when dry, ± rugose, lanceolate or ± lingulate, acute or obtuse, bordered by elongate cells, singly or doubly toothed; costa ending at or near the leaf apex, often toothed at back, bearing green lamellae on the upper surface; cells rounded-hexagonal, smooth; setae elongate, sometimes clustered; capsules cylindric, suberect, often curved; operculum long-rostrate; calyptra naked.

1. Leaves ± elliptic or obovate, broadly pointed,
   scarcely undulate; lamellae few or sometimes
   lacking, low, and inconspicuous . . . . . 1. *A. crispum*
1. Leaves lanceolate, acute, undulate; lamellae
   more numerous and better developed, conspicuous 2
   2. Lamellae up to 6 cells high, obscuring 1/7
      or less of the upper portion of the
      leaf when moist . . . . 2. *A. undulatum*
   2. Lamellae 7–9 cells high, obscuring 1/4
      or more of the leaf
      when moist . . . . 3. *A. angustatum*

1. *Atrichum crispum* (James) Sull. — Leaves elliptic to oblong-obovate, broadly pointed, not or slightly undulate; costa percurrent or shortly excurrent; lamellae few or sometimes lacking (up to 4), 1–3 cells high; setae single or occasionally in clusters of 2 or 3. —

[191]

On moist soil, often on banks of streams. Vermont to Michigan and Tennessee; British Columbia and Oregon; Europe.

ILLUSTRATIONS: Grout, *Mosses with Hand-Lens and Microscope*, Pl. 7 (as *Catharinea*). Jennings, *Mosses of Western Pennsylvania* (ed. 2), Pl. 64.

Upper Peninsula: Alger Co., *Steere*.

2. *Atrichum undulatum* (Hedw.) P.-Beauv. — Leaves oblong-lanceolate, acute, undulate; costa subpercurrent; lamellae 2–6, up to 6 cells high, obscuring 1/7 or less of the upper portion of the leaf when moist; setae single or rarely in clusters of 2 or 3. — On rich soil in woods and often at the wet margins of woodland brooks. Newfoundland to Alaska, south to California and South Carolina; Europe, North Africa, Azores, and Asia.

ILLUSTRATIONS: Conard, *How to Know the Mosses* (ed. 2), fig. 44. Grout, *Mosses with Hand-Lens and Microscope*, fig. 14a–b (as *Catharinea*). Jennings, *Mosses of Western Pennsylvania* (ed. 2), Pl. 30. Welch, *Mosses of Indiana*, fig. 21.

Upper Peninsula: Chippewa Co., *Thorpe*. Houghton Co., *Parmelee*. Keweenaw Co., *Povah*. Marquette Co., *Nichols*. Ontonagon Co., *Nichols & Steere*.
Lower Peninsula: Alpena Co., *Robinson & Wells*. Cheboygan Co., *Nichols*. Clinton Co., *Parmelee*. Eaton Co., *Darlington*. Washtenaw Co., *Kauffman*.

3. *Atrichum angustatum* (Brid.) BSG — Leaves narrowly oblong-lanceolate, bluntly acute, undulate; costa subpercurrent; lamellae 4–7, 7–9 cells high, obscuring 1/4 or more of the upper portion of the leaf when moist; setae usually single. — On light soil in dry, open woods, often on disturbed soil (of roadbanks, for example). Widespread in eastern North America; Europe and Asia.

ILLUSTRATIONS: Figure 145. Conard, *How to Know the Mosses* (ed. 2), fig. 43. Grout, *Mosses with Hand-Lens and Microscope*, fig. 14c (as *Catharinea*). Jennings, *Mosses of Western Pennsylvania* (ed. 2), Pl. 31. Welch, *Mosses of Indiana*, fig. 22.

Upper Peninsula: Houghton Co., *Parmelee*. Ontonagon Co., *Nichols & Steere*.
Lower Peninsula: Cass Co., *Darlington*. Cheboygan Co., *Nichols*. Clinton Co., *Marshall*. Eaton Co., *Parmelee*. Emmet Co., *Nichols*. Ingham, Kalamazoo, Leelanau, Monroe, and Oakland counties, *Darlington*. Washtenaw Co., *Kauffman*.

2. *Pogonatum* P.-Beauv.

Plants small to robust, scattered to loosely tufted; leaves rigid, larger above, scale-like toward the base, oblong-lanceolate from a wider, appressed base, acute or acuminate, usually toothed; costa wide, generally toothed at back above; both costa and lamina covered by numerous lamellae on the upper surface; cells rounded-hexagonal, smooth, incrassate; setae elongate; capsules cylindric, mostly erect; operculum rostrate; calyptra hairy.

1. Leaves long, linear-lanceolate, flexuous-contorted
when dry; terminal cells of lamellae ovoid or pyriform
in section . . . . 1. *P. alpinum*
1. Leaves short, lingulate or oblong-lanceolate, appressed
and scarcely altered on drying; terminal cells of
lamellae flat-topped in section . . . . . . 2. *P. capillare*

1. *Pogonatum alpinum* (Hedw.) Röhl. — Robust plants, usually branched; leaves linear-lanceolate; costa excurrent; terminal cells of lamellae papillose, ovoid or pyriform in section; capsules usually curved and inclined. — On soil. Greenland; arctic America south to New England, Colorado, and Washington; Europe, Asia, southern South America, and New Zealand.

ILLUSTRATIONS: Figure 146. Conard, *How to Know the Mosses* (ed. 2), fig. 46d. Grout, *Moss Flora of North America*, vol. 1, Pl. 60 (fig. 1–11), *Mosses with Hand-Lens and Microscope*, fig. 18c, c', c''.

Upper Peninsula: Alger Co., *Nichols*. Keweenaw Co., *Steere*. Marquette Co., *Nichols*. Ontonagon Co., *Nichols & Steere*.

2. *Pogonatum capillare* (Mx.) Brid. — Small, usually simple plants, naked and wiry below, densely foliate at the stem tips; leaves short and broad, imbricate and scarcely contorted when dry; costa shortly excurrent; terminal cells of lamellae papillose, quadrate to rectangular in section; capsules straight, erect or nearly so. — On soil of rocky banks or in crevices of cliffs. Greenland; across the continent, south to North Carolina and Oregon; Europe and Asia.

ILLUSTRATION: Grout, *Moss Flora of North America*, vol. 1, Pl. 59C.

Upper Peninsula: Alger Co., *Tinney*. Keweenaw Co., *Steere*.

3. *Polytrichum* Hedw.

Very similar to *Pogonatum;* capsules 4- or rarely 6-angled, with a sharply differentiated hypophysis at base, mostly inclined.

1. Leaves ending in a hyaline, toothed awn . . . 6. *P. piliferum*
1. Leaves not ending in a hyaline awn 2
  2. Leaf margins entire, abruptly inflexed above,
    covering the lamellae 3
  2. Leaf margins toothed, plane or erect,
    not inflexed 4
3. Plants of bogs; stems conspicuously
  tomentose . . . . . . 5a. *P. juniperinum* var. *alpestre*
3. Plants of drier habitats; stems not tomentose
  (except sometimes at base) . . . . . 5. *P. juniperinum*
  4. Terminal cells of lamellae notched . . . . 4. *P. commune*
  4. Terminal cells not notched 5

[193]

5. Terminal cells quadrate, flat-topped or slightly
   emarginate at top . . . . 3. *P. ohioense*
5. Terminal cells of lamellae rounded or ovoid,
   not much differentiated 6
   6. Leaves ± contorted when dry; lamellae not
   covering the entire lamina, leaving a noticeable
   margin 4–9 or more cells wide in the upper
   part of the leaf . . . . 2. *P. gracile*
   6. Leaves rigid and not much contorted when dry;
   lamellae occupying nearly all the lamina . . 1. *P. formosum*

1. *Polytrichum formosum* Hedw. — Plants fairly robust, rigid;
leaves not much contorted when dry; lamellae covering most of
the lamina, the terminal cells not much differentiated, rounded or
narrowly elliptic. — On soil and rocks. Greenland; eastern North
America south to North Carolina; Alaska to Oregon; Europe, North
Africa, Macaronesia, and Asia.

ILLUSTRATION: Grout, *Moss Flora of North America*, vol. 1, Pl. 61A.

Upper Peninsula: Marquette Co., *Nichols*. Ontonagon Co., *Nichols & Steere*.
Lower Peninsula: Cheboygan and Emmet counties, *Nichols*.

2. *Polytrichum gracile* Turn. — Very similar to *P. formosum;*
leaves somewhat contorted when dry; lamellae not covering the
entire lamina, leaving a noticeable, erect or incurved margin.

ILLUSTRATIONS: Grout, *Moss Flora of North America*, vol. 1, Pl. 61B, *Mosses
with Hand-Lens and Microscope*, fig. 16b. Jennings, *Mosses of Western Pennsylvania* (ed. 2), Pl. 72.

Upper Peninsula: Houghton Co., *Parmelee*.
Lower Peninsula: Kalamazoo Co., *Becker*.

3. *Polytrichum ohioense* Ren. & Card. — Coarse, rigid plants;
terminal cells of lamellae square or rectangular in section, flat-topped
or slightly notched. — On soil or humus in woods. Widespread in
eastern North America; Europe.

ILLUSTRATIONS: Conard, *How to Know the Mosses* (ed. 2), fig. 50e–g. Grout,
Mosses with Hand-Lens and Microscope, fig. 17e. Jennings, *Mosses of Western
Pennsylvania* (ed. 2), Pl. 32. Welch, *Mosses of Indiana*, fig. 26.

Upper Peninsula: Chippewa Co., *Thorpe*. Gogebic Co., *Bessey*. Keweenaw
Co., *Cooper*. Marquette Co., *Nichols*.
Lower Peninsula: Barry Co., *Gilly & Parmelee*. Cheboygan Co., *Ehlers*.
Clinton Co., *Parmelee*. Eaton and Ingham counties, *Darlington*. Kalamazoo Co.,
Gilly & Parmelee. Leelanau and Oakland counties, *Darlington*. Washtenaw Co.,
Kauffman.

4. *Polytrichum commune* Hedw. — Robust, rigid plants very
much like *P. ohioense;* terminal cells of lamellae broad and deeply

notched. — On soil in wet or moist woods. Greenland; arctic America, south to the Gulf of Mexico and Arizona; Europe, Africa, Macaronesia, Australia, and New Zealand.

ILLUSTRATIONS: Conard, *How to Know the Mosses* (ed. 2), fig. 33a–d. Grout, *Mosses with Hand-Lens and Microscope*, fig. 17a–d. Jennings, *Mosses of Western Pennsylvania* (ed. 2), Pl. 33. Welch, *Mosses of Indiana*, fig. 27.

Upper Peninsula: Chippewa Co., *Steere*. Houghton Co., *Mugford*. Keweenaw Co., *Allen & Stuntz*. Ontonagon Co., *Darlington*.

Lower Peninsula: Cheboygan Co., *Nichols*. Clinton Co., *Parmelee*. Ingham Co., *Beal*. Washtenaw Co., *Kauffman*.

5. *Polytrichum juniperinum* Hedw. — Plants coarse and robust; leaves entire with margins abruptly inflexed and covering the lamellae; costa excurrent as a red, toothed awn; terminal cells of lamellae in section ovoid-conic, very thick-walled. — On soil or rocks, usually in dry, open or partially shaded places, common on roadbanks. Greenland; almost throughout North America; nearly cosmopolitan.

ILLUSTRATIONS: Conard, *How to Know the Mosses* (ed. 2), fig. 48a–f. Grout, *Mosses with Hand-Lens and Microscope*, fig. 17o, o'. Jennings, *Mosses of Western Pennsylvania* (ed. 2), Pl. 33. Welch, *Mosses of Indiana*, fig. 29.

Upper Peninsula: Alger Co., *Wheeler*. Chippewa Co., *Thorpe*. Gogebic Co., *Bessey*. Houghton Co., *Parmelee*. Marquette Co., *Nichols*. Ontonagon Co., *Darlington*.

Lower Peninsula: Alpena and Cheboygan counties, *Wheeler*. Clinton Co., *Darlington*. Eaton Co., *Parmelee*. Ingham Co., *Wheeler*. Jackson Co., *Marshall*. Kalamazoo, Leelanau, Oakland, and St. Joseph counties, *Darlington*. Washtenaw Co., *Kauffman*.

5a. *Polytrichum juniperinum* var. *alpestre* BSG — Plants in dense tufts, matted nearly to the stem tips with tomentum. — Forming hummocks in peat bogs or wet coniferous forests, associated with *Sphagnum*. Greenland; arctic America to North Carolina, Wyoming, and Washington; Europe, Asia, southern South America, and antarctic regions.

ILLUSTRATIONS: Grout, *Moss Flora of North America*, vol. 1, Pl. 64B. Jennings, *Mosses of Western Pennsylvania* (ed. 2), Pl. 33. Welch, *Mosses of Indiana*, fig. 30.

Upper Peninsula: Chippewa Co., *Steere*. Delta Co., *Gilly & Parmelee*. Keweenaw Co., *Holt*. Schoolcraft Co., *Gilly & Parmelee*.

Lower Peninsula: Cheboygan Co., *Nichols*. Ingham Co., *Parmelee*. Washtenaw Co., *Steere*.

6. *Polytrichum piliferum* Hedw. — Plants coarse and rigid but usually low; leaves similar to those of *P. juniperinum* but ending in a hyaline, toothed hair-point. — On dry, often sandy or gravelly

soil, usually in the open. Greenland; across the continent, south to Ohio, Colorado and California; Europe, Macaronesia, Asia, and South America.

ILLUSTRATIONS: Figure 147. Conard, *How to Know the Mosses* (ed. 2), fig. 48g. Grout, *Mosses with Hand-Lens and Microscope*, Pl. 17g, h. Jennings, *Mosses of Western Pennsylvania* (ed. 2), Pl. 32 and 64. Welch, *Mosses of Indiana*, fig. 31.

Upper Peninsula: Baraga Co., *Parmelee.* Chippewa Co., *Steere.* Houghton Co., *Parmelee.* Keweenaw Co., *Allen & Stuntz.* Mackinac Co., *Wheeler.* Marquette Co., *Nichols.* Ontonagon Co., *Darlington.*

Lower Peninsula: Barry Co., *Gilly & Parmelee.* Cheboygan Co., *Nichols.* Eaton Co., *Parmelee.* Huron Co., *Kauffman.* Iosco Co., *Gilly & Parmelee.* Leelanau Co., *Darlington.* Washtenaw Co., *Steere.*

Glossary

Acicular, needle-like.

Acrocarpous, having the sporophyte or capsule terminal on the stem or a main branch.

Acumen, a narrowly tapered leaf tip (plural, *acumina*); *acuminate,* tapering to a slender point.

Acute, sharply pointed (less than 90°).

Alar cells, cells at the basal angles of the leaf, often differing in color, shape, or size from other basal cells.

Annulus, a ring of specialized, often enlarged and tumid cells between the mouth of the capsule and the operculum, serving in dehiscence.

Antheridium, the male reproductive organ, containing the sperms (plural, *antheridia*).

Apiculate, ending abruptly in a short tip (*apiculus*).

Appendiculate, having short, projecting spurs at intervals, usually used in describing the cilia of the inner peristome (see also *nodose*).

Appressed, lying close against the stem (mostly of leaves).

Archegonium, the female reproductive organ, a somewhat flask-shaped structure containing an egg cell (plural, *archegonia*).

Arcuate, arched, bent like a bow.

Areolation, the network formed by the cell outlines of a leaf.

Articulate, with thickened joints (of peristome teeth).

Ascending, rising more or less obliquely.

Astomous, without an operculum, applied to capsules lacking regular dehiscence.

Attenuate, slenderly tapering.

Auriculate, applied to leaves having ear-like lobes (*auricles*) at the basal angles.

Autoicous, having both male and female sex organs in separate buds on the same plant.

Awn, a bristle or hair-point, usually formed by an excurrent costa.

Axil, angle between leaf and stem; *axillary,* used in reference to buds or other organs in the leaf axils.

Bifid, 2-cleft, forked.

Bistratose, in 2 layers.

Bracts, specialized, often diminutive leaves surrounding the antheridia or archegonia.

Brood-bodies, reduced buds, leaves, or branches (*propagula*) or small, globose, ellipsoidal or cylindric, few-celled bodies (*gemmae*) which become detached from the plant and propagate new plants vegetatively.

Bulbiform, having more or less the form of a bulb; *bulbil,* a minute bulb-shaped propagulum.

Caespitose, tufted, forming cushions or sods.

Calcicolous, growing on a limey substratum (either soil or rock).

Calyptra, a membranous hood covering the developing capsule, derived from the upper part of the archegonium which continued growth after fertilization of the egg (plural, *calyptrae*).

Campanulate, bell-shaped (used to describe some mitrate calyptrae).

Canaliculate, channeled (as of the narrow part of a leaf with strongly infolded edges).

Cancellate, lattice-like.

Canescent, hoary because of hyaline hair-points.

Capillary, hair-like.

Capsule, the spore-case or "fruit" terminating the mature sporophyte; that part containing the spores is sometimes called the *urn* as contrasted with the lower, sterile portion often differentiated as a *neck.*

Carinate, keeled (often applied to the segments of the endostome and sometimes to leaves).

Castaneous, chestnut-brown.

Central strand, a group of elongate cells running centrally through the stems of some mosses.

Cernuous, nodding or drooping.

Chlorophyllose, green, containing chlorophyll (applied especially to the green cells forming the network in the leaves of *Sphagnum*).

Cilia, hair-like divisions of the inner peristome, when present, alternating with the segments (singular, *cilium*); *ciliate* is occasionally used to describe leaf margins or fringed calyptrae.

Circinate, approximating a circle in degree of curvature (sometimes applied to leaves).

Cladocarpous, sporophyte terminating a short archegonial branch.

Clavate, club-shaped, thickened upward.

Cleft, deeply split or lobed, usually to or below the middle.

Cleistocarpous, without an operculum (hence irregularly dehiscent).

Collenchymatous, thickened at the corners of cells.

Columella, central axis of the capsule around which the spores are produced (occasionally falling attached to the operculum).

Commissure, strictly, a seam (in *Sphagnum* the sides of the hyaline cells next to their union with the chlorophyll cells).

Comose, bearing a tuft (referring to leaves crowded at the tip of a stem).

Complanate, flattened.

Complicate, folded lengthwise.

Conduplicate, folded lengthwise along the middle (used of leaves).

Confluent, merging.

Constricted (or *contracted*), often applied to capsules becoming narrowed below the mouth on drying.

Convolute, rolled together and forming a sheath (used particularly in connection with perichaetial leaves having margins inrolled and clasping the base of the seta).

Cordate, heart-shaped.

Cortex, an outer, specialized layer of the stem.

Corticolous, on bark.

Cosmopolitan, generally distributed over the earth's surface.

Costa, the nerve of a leaf, sometimes double, sometimes single, forming a midrib (plural, *costae*).

Crenate, having rounded teeth; *crenulate,* finely crenate.

Cribrose, perforated with small holes, sieve-like.

Crisped, wavy, referring to leaves which are variously twisted and curved, especially when dry (as bacon becomes crisped on frying).

Cucullate, concave or hood-like at the apex (referring to leaves and calyptrae; a cucullate calyptra is slit on one side, resembling a monk's hood).

Cuneate, wedge-shaped.

Cuspidate, tipped with an abrupt, relatively stout point, as of some leaf apices.

Decumbent, ascending from a prostrate base (applied to stems).

Decurrent, with margins extending down the stem below the leaf insertion.

Defoliate, without leaves.

Dehiscent, splitting open (of capsules).

Deltoid, equilaterally triangular.

Dendroid, tree-like.

Dentate, having outwardly directed teeth; *denticulate,* finely dentate.

Dichotomous, equally forked (referring to branching).

Dimorphous, in 2 forms.

Dioicous, having the male and female sex organs on separate plants.

Distichous, inserted in 2 opposite rows.

Divisural line, the median, usually zig-zag line running the length of the peristome teeth.

Dorsal, the back (applied to the outer or lower surface of a leaf in contrast to the ventral portion which is directed toward the stem; also applied to the outer surface of a peristome tooth).

Ecostate, lacking a costa.

Emarginate, notched at the apex (usually used in connection with a leaf).

Emergent or *emersed,* applied to a capsule projecting slightly beyond the perichaetial leaves or only partially exposed (see also *immersed* and *exserted*).

Endostome, the inner peristome, usually consisting of segments, sometimes arising from a basal membrane, often alternating with cilia.

Equitant, straddling (like a rider on horseback), referring to leaves, especially in *Fissidens.*

Erose, irregularly notched (as if gnawed).

Excavate, abruptly concave (applied to the curved depression at the insertion or the basal angles of a leaf).

Excurrent, prolonged beyond the apex of the leaf (referring to the costa).

Exothecial, applied to the outer cells of the capsule wall (or *exothecium*).

Exserted, rising above surrounding parts, as a capsule projected beyond the perichaetial leaves because of an elongate seta.

Falcate, curved like a scythe; *falcate-secund,* curved and turned to one side.

Fascicle, a bundle or cluster of branches.

Fenestrate, perforated (with window-like openings).

Fibrils, the spiral wall thickenings found in the hyaline leaf or cortical cells of *Sphagnum; fibrillose,* lined with fine, thread-like structures.

Filiform, filamentous, thread-like.

Fimbriate, fringed.

Flaccid, soft, lax, or weak.

Flagella, small, whip-like structures, sometimes referring to slender, tapering stems or branches or to slender brood-branches (singular, *flagellum*).

Flexuous, bent alternately backward and forward (often applied to leaves, awns, or setae).

Flowers, the buds in which the reproductive organs are produced.

Foliate, having leaves.

Fruit, the ripened capsule.

Fugacious, falling away early.

Fulvous, brownish-yellow, tawny.

Fundus, bottom or lowest part, occasionally applied to the base of the peristome teeth.

Fuscous, dull-brown.

Fusiform, spindle-shaped, widest at the middle and tapered to the ends.

Gametophyte, the plant which bears the sexual organs; in mosses, the green leafy plants arising from the protonema.

Gemma, a small body of a few cells, capable of reproducing the plant (plural, *gemmae*); see also *brood-bodies* and *propagula.*

Geniculate, bent abruptly (like a knee).

Gibbous, swollen on one side (sometimes applied to capsules).

Glabrous, smooth.

Glaucous, whitened or bluish (resembling the waxy bloom on some fruits, such as plums).

Globose, spherical.

Granulose, covered with minute grains.

Gregarious, in groups, growing together but not in a tuft.

Guide cells, large, empty, rather thin-walled cells in a median row, distinguished from the small, thick-walled stereids seen in cross-section of the costa of some mosses.

Gymnostomous, lacking a peristome.

Habit, the general appearance of a plant.

Habitat, the environment of a plant.

Hastate, having more or less the shape of an arrowhead with the basal lobes pointing outward.

Hispid, beset with stiff hairs or bristles.

Homomallous, pointing in the same direction, secund.

Hyaline, transparent, colorless.

Hygroscopic, changing position with a change in humidity.

Hypnoid, with features similar to those of the genus *Hypnum* (often applied to a perfect, double peristome consisting of 16 lanceolate teeth and 16 keeled segments from a well-developed basal membrane and—typically —cilia alternating with them).

Hypophysis, an enlargement of the seta immediately below the capsule.

Imbricate, overlapping and appressed.

Immersed, applied to a capsule wholly concealed within the perichaetial leaves.

Inclined (capsules), more or less bent but not pendulous (usually less than 90° from the vertical).

Incrassate, thickened (referring to cell walls).

Incurved (leaf margins), more or less curved toward the midrib.

Indehiscent, not breaking open, operculum not differentiated.

Inflorescense, a cluster of reproductive organs, antheridia, archegonia, or both, and their surrounding leaves or bracts.

Infolded, abruptly turned inward (as the leaf margins of some species of *Polytrichum*.

Innovation, a branch formed below the terminal inflorescence after maturity of sex organs.

Insertion, point of attachment (as of a leaf on a stem).

Intricate, intertangled (used of densely matted plants).

Involute (leaf margins), inrolled.

Isodiametric, more or less equal in length and breadth.

Julaceous, slenderly cylindric and worm-like.

Keel, a sharp median ridge, especially that formed by the costa of a leaf; also applied to endostome segments.

Lacerate, irregularly slit into narrow divisions, as if torn.

Lamellae, thin ridges of green cells (characteristic of the upper leaf surfaces of certain genera, especially in the Polytrichaceae).

Lamina, the leaf blade.

Lanceolate, tapering from near the base to a narrow apex, as the head of a lance (narrower than ovate).

Lenticular, lens-shaped.

Lid, see *operculum.*

Ligulate, strap-shaped (narrow, elongate, and about the same width throughout).

Limb, the upper part of the leaf as contrasted with the broader or otherwise differentiated base.

Linear, long and narrow, with margins approximately parallel (narrower than ligulate).

Lingulate, tongue-shaped, oblong with the end somewhat broader and usually obtuse or rounded (see also *spatulate*).

Lumen, the cell cavity (plural, *lumina*).

Mammillose, having a nipple-like projection (as of the apex of some opercula or some bulging leaf cells).

Mat, a relatively thin mass of intertangled plants.

Median leaf cells, those from the central portion of the leaf as distinguished from those above or below.

Membrane gap, a portion of the wall of hyaline cells (in stem leaves of *Sphagnum*) which is lacking ("resorbed").

Micron (μ), .001 millimeter.

Midrib, the central nerve or costa of a leaf.

Mitrate, cap-like and divided into several equal lobes at base (referring to the calyptra); opposed to *cucullate,* which is slit well up from the base on one side.

Monoicous, with antheridia and archegonia on the same plant (see also *autoicous, paroicous, synoicous* and *polygamous*).

Mucro, a short, abrupt point; *mucronate,* ending in a short tip, usually formed by a shortly excurrent costa; *apiculate* usually indicates a somewhat longer tip and *cuspidate* a stouter one.

Muticous, without a hair point.

Neck (capsule), lower sterile portion next to the seta.

Nerve, see *costa.*

Nodose, marked by a series of knots or swellings along the length (of cilia of the inner peristome); *nodulose,* having small knots (see also *appendiculate*).

Ob-, a prefix indicating inversion, as in *obovate* (inversely egg-shaped, *i.e.,* broader at the apex than at the base).

Obtuse, blunt (more than 90°).

Operculum, the lid covering the mouth of the capsule and falling at maturity to allow the escape of spores (plural, *opercula*).

Orbicular, circular or nearly so.

Oval, short-elliptic.

Ovate, egg-shaped (with the narrower part above); *ovoid,* a solid with an ovate outline.

Papilla, a minute protuberance occurring especially on setae or leaf cells; *papillose,* roughened by papillae.

Paraphyllia, small, green, sometimes branched, filiform, lanceolate, or leaf-like scales on the stems or branches of some mosses.

Paraphyses, hairs growing among antheridia or archegonia (singular, *paraphysis*).

Parenchymatous, composed of rather thin-walled, usually short and undifferentiated cells.

Paroicous, monoicous with the antheridia surrounding the archegonia or in the axils of perichaetial or upper stem leaves.

Patent (usually of leaves), open or spreading, but less than 90°.

Pellucid, transparent or translucent.

Pendulous, pendent, drooping, cernuous (often applied to capsules).

[201]

Percurrent (costa), reaching the apex of the leaf but not beyond.

Perichaetial, referring to the female inflorescence (or *perichaetium*).

Perigonial, referring to the male inflorescence (or *perigonium*).

Peristome, the fringe of teeth around the mouth of the capsule, visible after the falling of the lid; it may consist of an outer circle of teeth (*exostome*) and an inner circle of (usually) segments (*endostome*).

Piliferous, bearing an awn or hair-point.

Pinnate, with numerous spreading, mostly equidistant branches arranged on both sides of the axis, often feather-like.

Pitted, referring to cell walls with small openings or depressions (also called *porose*).

Pleurocarpous, with the sporophyte terminating a specialized branch growing laterally from the main stem or along one of its prominent branches.

Plicate (mostly of leaves), folded longitudinally to form pleats.

Plumose, like a plume or feather (referring to very regularly pinnate branching).

Pluripapillose, with many papillae.

Polygamous, having antheridia and archegonia in separate buds and also mixed together on the same plant.

Pores, small openings or *pits* in the walls of some cells, especially conspicuous in the hyaline cells of *Sphagnum*, where they are sometimes ringed; *porose*, pitted.

Primary stem, the original or main stem, often prostrate and creeping, sometimes subterranean (see also *rhizome*).

Procumbent, prostrate, lying or trailing along the ground.

Proliferous, with young shoots arising among the bracts surrounding the sex organs.

Propagula, buds, reduced leaves, or tiny branches capable of reproducing the plant asexually (singular, *propagulum*).

Prosenchymatous (tissue), composed of narrow, elongate, overlapping cells with pointed ends.

Protonema, alga-like threads formed on the germination of a spore and giving rise to the leafy gametophytes (plural, *protonemata*).

Pseudopodium, a false seta, a projection of the stem tip in *Sphagnum* and *Andreaea* serving the function of a seta (plural, *pseudopodia*).

Punctate, marked with scattered dots.

Pyriform, pear-shaped.

Quadrate, square.

Radicles (see rhizoids); *radiculose*, covered with rhizoids.

Recurved, curved downward or backward.

Reflexed, bent backward.

Resorption furrow, a furrow produced in some species of *Sphagnum* by the disappearance or "resorption" of the outer wall of the cells at the leaf margins (seen in cross-section).

Reticulate, in the form of a network.

Retort cells, somewhat enlarged or elongate cortical cells in which the apex is curved outward and opens by a fairly conspicuous pore (found on the branches in many species of *Sphagnum*).

Revoluble, rolling away (applied to an annulus which falls in a ring).

Revolute (leaf margin), rolled backward (see also *reflexed* and *recurved*).

Rhizoid, a thread-like structure, dead at maturity, serving especially to anchor the base of the stem, sometimes forming a felt or *tomentum* on the stems (also called a *radicle*).

Rhizome, a slender, horizontal stem, usually below the surface of the ground, giving rise to secondary stems at intervals.

Rhomboidal, elongate-hexagonal; *rhombic* is shorter and diamond-shaped.

[202]

Rostrate, beaked (usually applied to the operculum).

Rugose, wrinkled crosswise (usually used to express extreme undulation).

Scabrous, rough, usually owing to papillae.

Scarious, membranaceous, scale-like.

Secondary stems, branches of the primary stems.

Secund, turned to one side (mostly of leaves).

Segments, the major divisions of the inner peristome (*endostome*).

Seriate, in rows.

Serrate, with teeth pointing forward; *doubly serrate,* with teeth themselves more or less serrate or in pairs; *serrulate,* finely *serrate.*

Sessile, without a seta.

Seta, the stalk bearing the capsule at its tip.

Setaceous, bristle-like.

Sinuous, wavy (applied especially to the shape of some leaf cells or cell walls).

Spatulate, gradually narrowed from a rounded summit (similar to lingulate but more abruptly broadened above).

Sphagniform, referring to enlarged, hyaline, porose cells, especially those of *Sphagnum.*

Spinulose, minutely spiny; *spinose* refers to coarse, sharp teeth or papillae.

Spores, minute, mostly spherical, nearly always unicellular bodies produced in the capsules and serving to propagate the moss plant.

Sporophyte, the spore-bearing generation, typically consisting of capsule, seta, and foot.

Squarrose, spreading at right angles; *squarrose-recurved,* horizontally spreading with recurved tips.

Stereids, thick-walled cells found in the costa of certain mosses.

Sterile, not fruiting.

Stolon, a runner, a slender, elongate stem or branch that is disposed to "root."

Stomata, pores usually found in the lower part of capsules, typically with 2 guard cells, similar in structure and function to those of higher plants (singular, *stoma*).

Striate, marked with fine longitudinal lines or ridges; *plicate, sulcate,* and *furrowed* or *ribbed* are more extreme conditions.

Strumose, a goiter-like swelling at the base of some capsules.

Sub-, a prefix signifying somewhat, nearly, or slightly, as *subglobose, subacute, subpercurrent,* etc.

Sulcate, more deeply grooved than *striate.*

Synoicous, with male and female sex organs in the same inflorescences.

Terete, circular in cross-section.

Tomentose, densely matted with rhizoids.

Tooth, a division of the outer peristome.

Trabecula, projecting cross-bar at the joints on the inner side of peristome teeth (plural, *trabeculae*).

Truncate (usually of leaves), with the apex abruptly cut off.

Tubulose, forming a tube because of incurved leaf margins.

Tumid, swollen or inflated.

Turbinate, top-shaped.

Turgid, swollen.

Uncinate, with the point curved, as a hook.

Undulate, wavy; *rugose* is a more extreme condition.

Unipapillose, with one papilla.

Unistratose, in one layer.

[203]

Urceolate, urn-like (applied to capsules which are broad at the mouth and abruptly narrowed at the neck).

Ventral, inner, applied to the surface of a leaf toward the stem (the upper surface) or the inner surface of a peristome tooth.
Ventricose, bulging on one side (like a stomach).
Vermicular, worm-like.

Whorl, arrangement in a circle around the stem.

References

ANDERSON, L. E. 1954. Hoyer's solution as a rapid permanent mounting medium for bryophytes. Bryol. 57: 242-244.

BRAITHWAITE, R. 1880-1905. The British Moss Flora. 3 vols. London.

BROTHERUS, V. F. 1923. Die Laubmoose Fennoskandias. 635 pp. Helsingfors.

CHOPRA, N. & A. J. SHARP. 1961. The occurrence of *Bryum coronatum* in Michigan. Bryol. 64: 64-65.

CONARD, H. S. 1962. *Pylaisia polyantha.* Bryol. 64: 355-363. (1961.)

————— & COMMITTEE OF THE SULLIVANT MOSS SOCIETY. 1945. The bryophyte herbarium. A moss collection; Preparation and care. *Ibid.* 48: 198-202.

CRUM, HOWARD. 1964. Mosses of the Douglas Lake Region of Michigan. Mich. Botanist, 3:3-12, 48-63.

DARLINGTON, H. T. 1938. Bryophytes collected in the vicinity of Glen Lake, Leelanau County, Michigan. Bryol. 41: 99-106.

DORE, W. G. 1953. Herbarium sheets for filing mosses. Bryol. 56: 297.

DUNHAM, ELIZABETH M. 1916. How to Know the Mosses. Ed. 1. 288 pp., 7 pls. Boston.

GRIFFIN, D. C. 1962. *Pohlia bulbifera*—a new record for Michigan. Bryol. 64: 383. (1961.)

GROUT, A. J. 1928-40. Moss Flora of North America North of Mexico. 3 vols. Newfane, Vt.

—————. 1947. Mosses with a Hand-Lens. Ed. 4. 344 pp., 81 pls. Newfane, Vt.

HERMANN, F. J. 1962. Additions to the bryophyte flora of Keweenaw County, Michigan. Rhodora 64: 121-125.

HILL, E. J. 1885. The Menominee Iron Region and its flora. Bot. Gaz. 10: 208-211, 225-229.

—————. 1902. *Fissidens grandifrons*, its habits and propagation. Bryol. 5: 56-58.

—————. 1909. Note on *Amblystegium noterophilum. Ibid.* 12: 108-109.

HOLT, W. P. 1909. Notes on the vegetation of Isle Royale, Michigan. Rep. Mich. Geol. Surv. 1909: 217-248.

HUTCHINSON, ETHEL P. 1954. Sectioning methods of moss leaves. Bryol. 57: 175-176.

KAUFFMAN, C. H. 1915. A preliminary list of the bryophytes of Michigan. Ann. Rep. Mich. Acad. Sci. 17: 217-223.

LIMPRICHT, K. G. 1890-95. Die Laubmoose Deutschlands, Oesterreichs und der Schweiz. 3 vols. Leipzig.

MAZZER, S. J., & A. S. SHARP. 1963. Some moss reports for Michigan. Bryol. 66: 68-69.

NICHOLS, G. E. 1922. The bryophytes of Michigan, with particular reference to the Douglas Lake region. Bryol. 25: 41-58.

—————. 1925. The bryophytes of Michigan, with particular reference to the Douglas Lake region. II. *Ibid.* 28: 73-75.

—————. 1932. Notes on Michigan bryophytes. *Ibid.* 35: 5-9.

—————. 1933. Notes on Michigan bryophytes. II. *Ibid.* 36: 69-78.

—————. 1935. The bryophytes of Michigan, with special reference to the Huron Mountain region. *Ibid.* 38: 11-19.

—————. 1937. The bryophytes of Michigan: A preliminary list. 18 pp. (mimeo.) New Haven, Conn.

————— & W. C. STEERE. 1937. Bryophytes of the Porcupine Mountains, Ontonagon County, Michigan. Pap. Mich. Acad. Sci., Arts, and Letters 22: 183-200. (1936.)

[205]

NYHOLM, ELSA. 1954. Illustrated Moss Flora of Fennoskandia. II Musci. Fasc. 1, pp. 1-87. Lund.

POVAH, A. H. 1929. Some non-vascular cryptogams from Vermilion, Chippewa County, Michigan. Pap. Mich. Acad. Sci., Arts, and Letters 9: 253-272. (1928.)

PRAEGER, W. E. 1920. A collection of *Sphagnum* from the Douglas Lake region, Cheboygan, Michigan. Ann Rep. Mich. Acad. Sci. 21: 337-388. (1919.)

ROBINSON, H. & J. WELLS. 1956. The bryophytes of certain limestone sinks in Alpena County, Michigan. Bryol. 59: 12-17.

RÖLL, J. 1897. Beiträge zur Moosflora von Nord-Amerika. Hedwigia 36: 41-66.

SAYRE, GENEVA. 1941. A gum arabic mounting medium for mosses. Bryol. 44: 160.

————. 1952. Key to the species of *Grimmia* in North America. *Ibid*. 55: 251-259.

SCHNOOBERGER, IRMA. 1940. Notes on bryophytes of central Michigan. Pap. Mich. Acad. Sci., Arts, and Letters 29: 101-106. (1939.)

SHARP, A. J. 1956. Factors in the distribution of *Hyophila tortula* and an extension of its known range to include Michigan. Mitt. Thüring. Bot. Gesell. 1 (2/3): 222-224.

STEERE, W. C. 1931. Notes on the mosses of southern Michigan. Bryol. 34: 1-5.

————. 1933. Notes on the mosses of southern Michigan. II. *Ibid*. 36: 24.

————. 1934. The bryophytes of the Chase S. Osborn Preserve of the University of Michigan, Sugar Island, Chippewa County, Michigan. Amer. Midl. Nat. 15: 761-769.

————. 1937. Critical bryophytes from the Keweenaw Peninsula, Michigan. Rhodora 39: 1-14, 33-46.

————. 1938. Critical bryophytes from the Keweenaw Peninsula, Michigan, II. Ann. Bryol. 11: 145-152.

————. 1942. Notes on Michigan bryophytes. IV. Bryol. 45: 153-172.

————. 1947. The bryophyte flora of Michigan. Pap. Mich. Acad. Sci., Arts, and Letters 31: 34-56. (1945.)

SULLIVANT, W. S. 1864. Icones Muscorum. 216 pp., 128 pls. Cambridge, Mass.

————. 1874. *Idem*. Supplement. 106 pp., 81 pls. Cambridge, Mass.

THORPE, FRANCES J. & A. H. POVAH. 1935. The bryophytes of Isle Royale, Lake Superior. Bryol. 38: 32-46.

WYNNE, FRANCES E. 1942. Studies in *Drepanocladus* in North America north of Mexico. Ph.D. dissertation. University of Michigan, Ann Arbor.

————. 1945. Studies in *Drepanocladus*. IV. Taxonomy. Bryol. 47: 147-189. (1944.)

[206]

Index

[208]

[209]

[211]